Delhi
A Nature Journal

'*Delhi: A Nature Journal* by Anuradha Kumar-Jain, beautifully illustrated by Bahaar Meera Jain, captures the evocative seasons of Delhi, intertwining its urban landscape with the natural world. This journal invites readers to connect with the rhythms of nature, showcasing the city's vibrant ecosystems—its birds, plants, flowers, changing skies, animals and water bodies. Through stunning visuals and reflective prose, it offers both an escape and a reminder, encouraging a slower pace, in harmony with nature's cycles. In an era of critical climate action, this journal is a vital reminder of the importance of preserving our natural environment.'

—**Amitabh Kant**
G20 Sherpa, Government of India, and
Former CEO, NITI Aayog

'Anuradha's eyes become the windows to Delhi's soul in the remarkably written and beautifully illustrated *Delhi: A Nature Journal*. Delhi rarely gets recognized for the treasure trove of its natural beauty, in the form of its sprawling nature, archaeological parks and flourishing tree cover. The author has made it possible to look at the unseen delights of the city that ironically hide in plain sight.

'The fascinating diversity of the species in Delhi, their behaviour and the interesting patterns of their interactions are well presented. They hold valuable lessons on nature's harmony to make things work. The beautiful and vivid descriptions of the natural world captured by the author invoke a feeling of 'Biophilia', the inherent human bond with nature comes alive amidst busy urban life. This book serves as a great example of how urban biodiversity and its conservation can be part of modern society. *Delhi: A Nature Journal* is a real inspiration to take a moment every day and immerse in the pleasures of the natural world that surrounds us.'

—**Marisa Gerards**
Ambassador of the Kingdom of the Netherlands to India, Nepal and Bhutan

'Through this book, the author captures a worldview of Nature, its many colours and forms, albeit in an urban setting of our ancient city. With the creativity of a writer/artist, the book is a visualization of our vibrant and rich natural heritage and a reading takes one through the days and seasons of Delhi and its surroundings. We need books like this to remind us of the natural world—the songs of the Oriental magpie-robin and changing tree phenology, the colours of the season, the historical settings, the interconnection among people. Of the myriad writings on Delhi through the years, this book is a creative contribution.'

—**Ravi Singh**
Secretary General & CEO and
(Ex-Officio) Trustee, WWF-India

Also by Anuradha Kumar-Jain

Written on the Wind

Delhi
A Nature Journal

Anuradha Kumar-Jain

Illustrations by Bahaar Meera Jain

Published by
Rupa Publications India Pvt. Ltd 2024
7/16, Ansari Road, Daryaganj
New Delhi 110002

Sales centres:
Bengaluru Chennai Hyderabad
Jaipur Kathmandu Kolkata
Mumbai Prayagraj

Copyright © Anuradha Kumar-Jain 2024
Illustrations © Bahaar Meera Jain

The views and opinions expressed in this book are the author's own and the facts are as reported by her, which have been verified to the extent possible, and the publishers are not in any way liable for the same.

All rights reserved.
No part of this publication may be reproduced, transmitted or stored in a retrieval system, in any form or by any means, electronic, mechanical, photocopying, recording or otherwise, without the prior permission of the publisher.

P-ISBN: 978-93-6156-369-0
E-ISBN: 978-93-6156-240-2

First impression 2024

10 9 8 7 6 5 4 3 2 1

The moral right of the author has been asserted.

Printed in India

This book is sold subject to the condition that it shall not, by way of trade or otherwise, be lent, resold, hired out or otherwise circulated, without the publisher's prior consent, in any form of binding or cover other than that in which it is published.

For my two passionate naturalists
Bahaar and Maahi

Contents

Introduction / ix

January / 1

February / 24

March / 41

April / 65

May / 83

June / 101

July / 117

August / 133

September / 147

October / 159

November / 177

December / 193

List of Trees / 209

References / 215

Introduction

'For nature gives to every time and season some beauties of its own; and from morning to night, as from the cradle to the grave, is but a succession of changes so gentle and easy that we can scarcely mark their progress.'

—CHARLES DICKENS, Nicholas Nickleby

Essentially a nature journal of Delhi, this book attempts to record the changing seasons in an urban landscape; the flora and fauna, the moments of transition, the sky and clouds, the ever-shifting weather patterns. Every season paints the land in multi-hued abstractness, much as a dyer colours each fabric differently, and to be aware of these subtle changes is to make every day more rewarding. It is very easy, in a world geared to work and rush and more work, to lose sight of the beauty surrounding us. Appreciating nature creates pockets of calm in our busy everyday lives, and is a gift to be cherished. The serendipitous glimpse of a colourful flower growing in an unexpected place or hearing the dawn chorus while still lying in bed are balm for the soul.

*'There's music in the sighing of a reed;
There's music in the gushing of a rill;
There's music in all things, if men had ears:
Their earth is but an echo of the spheres.'*

—LORD BYRON, Don Juan, Canto 15

Contrary to popular thought, an urban landscape, even one as congested as Delhi, can become a nature lover's delight. The trick is to choose what you see. A thing as simple as noticing the clouds scudding overhead and listening to the barbet in the trees by the roadside can make the endless traffic jams, which are such an inescapable part of life in Delhi, bearable, and maybe even interesting. Life thrives everywhere; hydrothermal vents on the ocean floor, where temperatures can reach upwards of 300°C, have entire vent-bio communities consisting of microbes and clams, snails and fish. In the Antarctic, penguins live in temperatures of −60°C on the sea ice. In comparison, Delhi is a cakewalk! It teems with life; its tree-lined avenues, parks, wetlands and private gardens support an amazing diversity of creatures and plants.

Situated on the banks of the river Yamuna, Delhi lies 28.24°–28.53° north and 76.50°–77.20° east, at an altitude of just 709 feet above sea level.[1] It is bordered by the Indo-Gangetic plains in the north and east, the Thar Desert to the west and the Aravallis to the south. The tropical location, coupled with the fact that the city is completely landlocked, has greatly influenced its climate, as has its proximity both to the Himalayas and the Thar Desert. Because of this, Delhi has a paradoxical mix of humid subtropical monsoon (Köppen *Cwa*) and semi-arid (Köppen *Bsh*) climatic types, with scanty rainfall (annually about 750 mm or 31 inches) and huge extremes of temperature.

Delhi has five distinct seasons: winter, spring, summer, monsoon and autumn, although the spring and autumn are short-lived. Summers, which extend from April to June, are dry and scorching, with an average temperature of 38°C, which can go up to 46°. Loo, an extremely hot and dry wind, originating from the Thar Desert, blows across the plains of North India, and afternoons are often blighted by thunderstorms. Monsoons, which last from July till the end of September, do bring some relief to the parched city, but the interregnum between showers can be uncomfortably humid. By October, the days start getting pleasant, and the average temperature for the month is between 19°C and 33°C. Winter, which lasts from November-end to about the middle of February is the best time to be in Delhi, although late December and early January can be particularly cold, with the temperature often dropping to single digits. This season is followed by a short but sweet spring when the gardens and parks are full of flowers.

'Live in each season as it passes; breathe the air, drink the drink, taste the fruit, and resign yourself to the influence of each.'

—HENRY DAVID THOREAU, Walden

Delhi has been the seat of imperial power for the last several hundred years. The Mahabharata refers to it as Indraprastha, a city established by the Pandavas between 600 and 1500 BCE. During the course of its long journey spanning millennia, the city has reinvented itself continuously, having been destroyed and rebuilt seven times at various neighbouring locations.[2] It has witnessed silently the rise and fall of empires, the intermingling of different faiths and cultures, grandeur and penury, and each one of these influences has contributed profoundly to the unique flavour of present-day New Delhi. It is a beautiful city, dotted with imposing forts and palaces, religious monuments, museums and other essentials of a bustling metropolis, which is also the seat of the national government.

The topography of Delhi consists of two distinct zones: the Yamuna floodplains and the Ridge. The latter is part of India's oldest mountain range, the Aravallis, thought to have formed about 2.5 billion years ago. Traversing 670 km in the south-west direction, the Aravallis extend diagonally through Gujarat to Rajasthan to Haryana, before culminating in Delhi. Entering the city from the south through Gurgaon (officially Gurugram), the Aravallis fork into two—one segment running down to the Yamuna through Mehrauli, and the other crossing Tughlaqabad into Dhaula Kuan in South Delhi. The southern part of the ridge comprises a continuous stretch of native forest and includes the Asola-Bhatti Wildlife Sanctuary, Mangar Bani, Bhondsi and Damdama. This is scrub and dry deciduous forest, also called tropical thorn forest. Common trees here are acacias and mimosas, like the babool, ronjh and khair. Native trees include dhau, hingot, khair, dhak, kareel and kumttha, many of which have vanished from other parts of Delhi.[3]

The geographical location and topography of Delhi have blessed it with distinct micro-habitats: the ridge, or Kohi; stretches of old alluvium, or Bangar; Khadar, or the new alluvium of the plains of the Yamuna; and the Yamuna floodplain itself, also called the Dabar. Most of this land has been built upon, often with disastrous consequences, but there still remain pockets of greenery interspersed with concrete, and these are home to an amazing variety of flora and fauna. The city has upwards of a thousand water bodies; while many are little more than sewage and effluent discharge ponds, many more support several resident and migratory waterbirds. Delhi also has nearly 18,000 parks and gardens, spread over 8,000 ha.[4] These include green areas and wide roadside verges, as well as big parks, which are home to a huge diversity of trees, including some relics of old forests. Among the important parks are Sanjay Van, Nehru Park, Lodhi Garden, Sunder Nursery, Garden of Five Senses and many others; together, Delhi's green areas support nearly 250 species of trees, which are true mirrors of the changing seasons, reflecting the passage of the year in their life cycle.

January

January is a beautiful month in upper India. At its beginning the days are short, for the sun does not rise until a quarter past seven and it sets again by half past five; but as the weeks pass, daylight gradually lengthens. Usually the sky is blue and the sun smiles benevolently. […] Delhi's gardens are then full of flowers. The varied lovely hues of blooms along herbaceous borders and other flower-beds combine with extravagant splashes of colour on bougainvillaea, poinsettia, bignonia, and varied shrubs and creepers, to make a gay show.[1]

1st

The first month of the Gregorian calendar, named after the Roman God Janus, is a wonderful time of year in Delhi. And whilst some days are beset by dense fog, which does not lift until the afternoon, the overarching mood is one of lightness. The year is new, spring is just around the corner, flowers spill out of containers and garden beds, and the air is cold and invigorating. The weather remains very pleasant, with an average high of 19°C and low of 8°C. Western disturbances often bring some rainfall, and I welcome the variety this brings to the days; sunny, with azure skies one moment, and moody grey the next. The days are short, with late sunrises and early sunsets, but towards the end of the month, the evenings start feeling considerably longer.

> There is something odd about the length of days and nights. […] On New Year's day the sun rises at 7.14 a.m. On the following day, instead of rising earlier, it can be later by a minute. At the other end, it sets at 5.35 p.m. on the first two evenings, thus adding a minute a day. By Republic Day (26th January) while it rises only three minutes earlier in the mornings, it sets as many as 20 minutes later (5.55 p.m.). Apparently this eccentricity is due to the elliptical shape of the earth.[2]

The year started with an unusual sighting of a leucistic laughing dove just outside my bedroom window. It was feeding with a small flock of other doves, and it took me a couple of minutes to overcome my confusion and identify it. I rushed to get the camera, but it flew off before I could take a photograph.

2nd

Wanted to usher in the New Year with style, and since the weather forecast seemed promising, we braved the morning fog to drive down to the Sultanpur National Park, approximately 50 km from Delhi, to see the winter migrants. Delhi has upwards of 450 species of birds ever sighted, both migrants and residents, although, on a single day, about 250 sightings is what the city reports, making it one of the most avian-friendly cities of the world (these numbers are for NCR).[3] The Big Bird Day in 2020 had a count of 253 species, less than the high of 271 in 2005, but more than the 247 counted in 2019. In 2021, the number again fell to 244.[4] Delhi and the rest of India lie on the Central Asian Flyway, one of the nine flyways of the world used by migrating waterbirds, with India alone supporting more than 250 species of migratory waterfowl in its numerous wetlands and nature parks.

Sultanpur was a delight. The fog had lifted by the time we reached, making the sighting of birds much easier. We were greeted at the entrance by a red-breasted flycatcher, a winter visitor to our corner of the world from its breeding grounds in Eastern Europe and Central Asia. Perched on the dried stump of an old tree, the bird made up in pugnaciousness what it lacked in size, conducting skilful aerial sallies to catch its prey. A short distance down the lane, we saw another winter migrant, the black redstart, a smart little bird with the habit of constantly flicking and vibrating its tail.

The Sultanpur Jheel attracts a huge number of migratory waterbirds. In the space of a couple of hours, we saw black-winged stilts, Eurasian teals, spot-billed pelicans, wood and spotted sandpipers, northern shovelers, pintails, woolly- and black-necked storks, greylag and bar-headed geese and a host of other migratory and resident waterbirds dotted around the lake. In the marshy land along the margins were purple swamphens and common moorhens and at least two types of kingfishers. On the verges next to the road were yellow and white wagtails, and my favourite of all, a bluethroat, a winter visitor from the Palaearctic region.

All in all, a very rewarding start to the new year!

3rd

A very foggy start to the day. Listening to the rain pelting down on to the windowpane, I snuggled a little deeper into my quilt. It was early morning, and the louring clouds made the dark linger a little longer. These are my favourite kinds of mornings; there is something festive about switching on the lights on a cold winter's day, while it is still slightly dark outside. Such days, especially if the showers continue intermittently, usually end with a dense fog settling in towards the evening. I love the hush of the fog: the cocooning silence, the white blanket obscuring the world and the birdsong that starts up almost immediately after it lifts. I especially enjoy watching the leaden skies from the warmth of my bed, with its wonderfully soft, hand-knitted throw.

Pansies have been the only spots of colour on an otherwise dreary day. The last few years have seen a huge increase in the size and variety of pansies available in the nurseries. I have deep pink and light mauve ones and maroon ones with yellow and black centres. But somehow, I still love the smaller purple and yellow cat-faced ones the best. As a child I remember thinking that each flower had its own mood, and in my mother's garden where they grew in glorious abandon, I would try to guess the attitude of each. So we had an angry cat, a smiling one and one with a beard growing next to a family of scowling felines. Violas are the smaller and more delicate cousins of pansies, and I always marvel at how perfectly formed each tiny flower is and how a pot bursting with a profusion of these little jewels can outshine much bigger and gaudier blooms. My current favourites are the tiny purple ones, so dark as to appear almost black, and the slightly bigger deep orange ones, which have been putting up a grand show for more than a month now. The nicest thing about pansies is their long flowering season, which lasts in Delhi from late November to mid-March, and it takes only regular deadheading and the occasional feeding to keep them in continuous bloom.

4th

I often wonder at the magic at work in nature; the wind blows, the sun shines, clouds drift across the sky, a butterfly dances by, the flowers in my garden bloom one by one and my favourite tailorbird trills sweetly from a bush nearby. All this at its appointed hour, without fuss or noise or any attention seeking. Sweet peas perfume the air, while the basil flavours our food, and as I brush past the young lemon tree growing in a corner in a big pot, I find its heady fragrance clinging to me.

A lovely cold, crisp day; bright and sunny but with a chill wind. The kind of day that immediately reminds me of winters from long ago, when the beautiful azure of the sky was belied by the frozen winds that blew down from the mountains. I have hung bird feeders outside the L-shaped window of my bedroom; three terracotta, two wooden hut-shaped and two bottle ones. Recently, I added a small bird table in the same corner of the garden, which is actually a recycled study table that the kids used when they were younger. The terracotta feeders have been topped up by a mixture of millets and sunflower seeds, groundnuts and the occasional red chilli for the rose-ringed parakeets. The smaller feeders that are frequented by the silverbills, sparrows and tailorbirds have a mixture of bajra and millets, while the long bottle feeders, which have perches for the birds to balance on, have a sparrow feed that I buy ready-mixed. However, I have noticed that the rose-ringed parakeets are partial to the bottle feeders, even though they have to do considerable manoeuvring to be able to reach the seeds with their beaks. They behave like small acrobats, tumbling and balancing and tumbling again, all the while squabbling among themselves, and eventually, the losers are banished to the terracotta pots, which incidentally have their favourite food!

I have also hung two terracotta water bowls in a corner of the garden near the feeders; the bigger one is from a small jatropha tree, which has leathery oval leaves and bears small flowers in a very pretty shade of red.

5th

Very cold and foggy and even the Persian lilac tree, growing just a few feet away, is no longer visible. But I like the fog, like its warm enveloping cocoon and the way everything within the wall of white is accentuated. My poinsettias are glowing jewel-like, the orange and mauve primulas are beginning to bloom and the ranunculus are doing great. The pansies are already putting up a great show, while the pinks (dianthus) and the stocks are in bud. Nasturtiums have started clambering over the small round box balls that I have fashioned so painstakingly and will have to be disciplined. Unfortunately I have had very little success with my favourite sweet peas; twice I have planted the seeds, but with hardly any luck. The larkspurs and lupins that I bought as seedlings are doing better, but I am not very happy with the antirrhinums or dog flowers, which are thin and sickly looking. In contrast, the daisies are growing into healthy plants and I peer at them closely every day, waiting for the first buds to emerge. Petunias are great for early colour in the garden, and mine are already filling their baskets. My favourites are the bi-coloured ones; white stripes on pink or pink on white or red and white or purple and white or purple star-shaped. This year I also have a beautiful dark purple that appears almost black. The hibiscuses have all had to be pruned drastically because of an attack by mites. My daughter insists on keeping them all grouped together, which I think is a bad idea. It just leads to all of them becoming afflicted at the same time.

6th

Our garden gets the winter sun for a large part of the day, from around 10 in the morning to half past four in the afternoon, and any time spent there, toasting frozen toes, is my idea of heaven. Add a bit of breeze and you have the heady combination of cold and sun and wind, which makes for my favourite kind of day, a day on which anything seems possible, when a hot cup of coffee cradled between cold hands can inspire an entire poem and when even the argumentative notes of the myna can sound melodious.

Can light be wind-blown? On a day like today, I like to imagine that it can, for it is only when the wind cavorts with the clothes hanging on the washing line to dry, pushing them out of the way, that the sun gets in, warming the pots of geraniums on the window sill. And later, when the wind, still frisky, sifts through the pile of dried leaves swept into a heap under the peepal tree by the gardener, dislodging them and scattering them about like rudderless boats adrift in a tumultuous sea, the grass growing between the cobbles gets its first glimpse of daylight.

I sometimes wish it could always be like this; cold day, warm sun, playful wind, bees hovering over the daisies, peepal leaves conducting an oratorio, the mellow timelessness of the afternoon stretching into eternity. But then I think of other things that I love; clouds scudding across the sky, the pitter-patter of raindrops, the first mogra to bloom in the scorching heat of summer; and I am grateful for the changing seasons, for winter turning into spring, for monsoon clouds gathering on the horizon, for the birds that migrate down from their homes in the high reaches of the Himalayas.

7th

Dark, overcast and cold, and the lights were on in the café at the top of the stairs when I reached mid-morning. Moody jazz playing in the background and a flavourful latte. Recipe for happiness! Later in the day a weak sun filtered through the leaves of the peepal, falling on the grey cobbles like tiny pools of gold.

Back home, I shifted to the best place to be on a cold winter's day—a day bed right next to a largish window. Or rather a bay-bed, in a small alcove off the living room, made cosy by cushions and colourful hand-knitted throws. From here, I can observe clouds chasing each other across the sky, the wind playing with the leaves of the neem, the effortless glide of eagles far overhead and the antics of drongos with whom I share a fondness for grey, foggy days.

8th

It has been raining intermittently since last night, and the fog, not having lifted even at midday, hangs low, reducing visibility and turning the outside a uniform shade of brownish grey. It's an eerie feeling, this unvariedness of the hours, windless and slate coloured from morning to evening, with no afternoon spike in brightness and hardly any birdsong, except for the 'tchee-tchee tchee' of the resolute drongos as they go about their business. Long, dreary days.

> *'People close their windows tight, light fires,*
> *keep warm in the sun and wear heavy garments: [...]*
> *Cold, cold, with heavy dews falling thick,*
> *and colder yet with the moonbeam's icy glitter,*
> *lit with ethereal beauty by wan stars,*
> *these nights give no comfort or joy to people.'*
>
> —KALIDASA, *Rtusamharam*, Canto V: Winter

9th

A late afternoon in winter, and the sun is beginning to draw in its rays, and shines now only on the trees in the distance, having retreated from where I am sitting. And suddenly the part that is sunlit seems more alive, almost festive in its golden glow. Some things, however insignificant they may seem, are forever yoked to memory, and with a power disproportionate to their importance otherwise, have the ability to transport a person to places far away and events a lifetime ago. In my parents' house in Chandigarh, I remember on lazy winter afternoons—when the air was redolent with the fragrance of sweet peas—chasing the sun across the length of the verandah, dragging my chair onto the patch of sunlight still remaining, wanting to extract the last amount of warmth on a cold day. My memory then takes me to Chamba in Himachal Pradesh, to the British-era circuit house, with its time-worn wooden eaves under which the sparrows nested. The upstairs rooms opened at the back onto a covered verandah overlooking the Ravi, and the song of the river was a mellifluous constant all through the day and night. Lazing on one of the planter chairs after a long day spent traversing the steep streets of the town, ruminating, enjoying the approaching dusk, savouring the very last drops of the flavourful elaichi chai, I would sit there until the day turned into night. The sun withdrew gradually, bidding farewell to my corner of the verandah first, travelling leisurely downslope to the bank of the river, pausing there momentarily to glint golden on the water, before crossing over to the mountains on the other side, from where it dissolved into the twilight in a blaze of glory. I remember thinking what a festive procession the sun and its retinue of luminescence made, the light radiant rather than dazzling at this hour.

10th

Intermittent rain the whole of last week, which has washed clean every surface, every leaf and bud, made the very air sparkle. It has also stripped the Persian lilac tree bare, with the yellowing leaves and coffee-coloured berries forming an intricately patterned carpet on the driveway.

The days are beautiful; cold and crisp or windy and sunny, but always exhilaratingly fresh. Golden sunshine, when it finally reaches the garden around noon, conjures up butterflies out of thin air, just like a magician pulls long-eared bunnies out of his hat. Yellow orange tip, common gull, common emigrant, common grass yellow, red Pierrot, plain tiger, blue tiger, cabbage white, common Mormon, all flit about dancing from flower to flower. So many flutterbies!

Sunlight and a lull in the rain also bring the birds out in hordes. A flock of Oriental white-eyes on the saptaparni, foraging at mid-level, constantly keeping up their soft call, reminds me of bells ringing in the distance. Another flock of silverbills is at the feeders. They are adorable, but rather silly little birds, who take the adage 'follow thy leader' to new ridiculous heights, all congregating on the same feeder, falling over themselves, even though the adjoining ones are empty. Then one will fly to the terracotta water pot for a refreshing sip, and all will follow till they totally encircle the vessel. If a member of the flock decides to feed on the ground, all the rest want to feed on the ground too. Maybe it has something to do with safety in numbers. But these little finches certainly have pluck, and will not hesitate to feed with much bigger birds.

I have noticed that the smaller birds—tailorbirds, silverbills and even sparrows—do not wait for other bigger birds to put in an appearance first before they start feeding.

On the other hand, mid-sized birds like bulbuls and magpie-robins, mynas and doves will wait for other members of their tribe before they fly down. One of the pleasures of the bird table this season is the flock of more than 20 red-whiskered bulbuls who come to feed on most days. However, they are lamentably timid and will wait for the rush to be over before they venture down from the adjoining bushes. Even then, the first to arrive will hesitate and look around for some seconds before picking up a morsel. However, almost all bulbuls are partial to the orange halves I put out for them and will chase away friends and relatives aggressively. The babblers, who are always in a large group, do not need any reassurance and will visit whenever they feel like it, but their palate seems to be somewhat adventurous and they like a little variety in their food. Maybe that is why they never visit the bottle feeders, which only have a mixture of grains and seeds. The Oriental white-eyes are spirited little creatures, who live in a world of their own, never drawn to the feeders, and never daunted by my presence nearby. A small flock will visit the water pot every day late in the afternoon and will depart as soon as its thirst is quenched.

11th

A family of spotted owlets have their nest on the goolar tree adjoining the side wall of our house and provide almost daily entertainment. I say family because three or four of them live in the same hollow on a thick side branch of the tree. We are careful to not go very close to the tree because of fear of alarming them, and also because they are our neighbours and we guard their privacy very zealously!

What a delight they are, bobbing their heads up and down at the sight of any intruders, snoozing often with one eye open and the other shut (my son has a photograph in which they are doing just that!), or sitting cheek by jowl on a branch nearby, watching the world go by. At dusk they decamp for the street lights, probably in a bid to procure dinner, and their distinctive hooting can be heard through the night and well into daybreak.

> Spotted owlets are pretty creatures. […] When unexpectedly disturbed, they behave in an odd manner,

shrieking rudely at the intruder, bobbing up and down on their perches, and glowering threateningly at him in an apparent attempt to frighten him away.

After the first few weeks of our acquaintance, they became used to me and rarely treated me to these demonstrations of disapproval. Nevertheless, whenever I bobbed my head up and down at them they at once returned the compliment by bobbing theirs up and down at me. But they no longer accompanied this action with vocal protests. The gestures seemed to be just a friendly exchange of diplomatic courtesies.[5]

13th

Today is Lohri, one of my favourite festivals. Maybe I am prejudiced because it is my son's birthday as well or maybe because I absolutely love the gajjak and rewri and peanuts that are such an integral part of the celebrations! Falling one day before Makar Sankranti, Lohri traditionally commemorates the passing of the winter solstice and the gradual lengthening of the days. In Punjab, where the winters are really cold, the celebrations include lighting a bonfire and sitting around it long into the night. It is an ancient festival and part of the folklore of the region.

Tomorrow is Makar Sankranti, the day the sun enters the Capricorn zodiac. Also called Uttarayan or Maghi, it is one of the very few festivals in India based on the solar calendar, which is why it falls on the same date every year. Uttarayan means northward movement, and like Lohri, Makar Sankranti celebrates the passing of the month with the shortest days. The northward movement of the sun signals a shift in seasons and a return of longer, warmer days. Makar Sankranti is also celebrated as a harvest festival in most parts of the country; the reaping is over and the new crop has been sown, leaving people with some leisure time to enjoy, although each state celebrates it with its own distinct traditions.

14th

Another bright sunny day and birding activity is very upbeat, with some additions to our usual friends; three brahminy starlings, a pair of treepies and a lesser whitethroat, although the latter never ventured down from his perch on the overhanging branch of the saptparni.

The small mandarin orange tree in the corner of the garden near the water pot has been attracting large green barbets ever since the fruit started ripening. A pair appears about noon every day and after settling on a fruit-laden branch, proceeds to gobble up the tiny oranges, tugging incessantly on the more stubborn ones. They are rather confiding birds and don't mind my presence a few feet away. Today, one of them, probably encouraged by the sun being out, decided to take a bath in the larger of the terracotta pots. It was a most amusing sight; the bird going round in circles, splashing about with its beak, but hesitating to lower itself into the cold water!

15th

Our local nursery is a patchwork of colour in the winters and reminds me of the blankets my gran used to crochet, the ones in which each small square was of two different hues, with the inner one usually in a more vibrant shade than the border. Rows upon rows of potted flowers perfume the air with their fragrance, and I spent a delightful morning choosing colourful pansies in tiny terracotta pots, wine-coloured verbena and orange nasturtiums for the hanging baskets and gorgeous pink daisies for my cup planters. There are so many flowers—double-flowered stocks, marigolds and calendulas, petunias and primulas, cinerarias and salvias, baskets of candytuft and alyssum—that it becomes difficult to choose which ones to buy, but since the children particularly wanted dahlias this time, we got a couple of dwarf ones and also two more ranunculus—one white and the other a beautiful dusky pink. I had some empty baskets at home, so couldn't resist a couple of double-striped petunias as well.

'Earth laughs in flowers.'

—RALPH WALDO EMERSON, Hamatreya

16th

An atmospheric day; cold, windy, overcast, with the sun struggling to peep out from behind the clouds. Perfect for browsing in old bookshops and drinking dark aromatic coffee; a moody brew for a moody morning.

The leaves of the jarul are now tinged with warm, woody shades of russet and brown, and from a distance the tree looks to be a uniform mahogany in colour; the bark, fruit and leaves all tinted with the same hue. Another tree that catches my attention throughout the year is the pilkhan; in spring with its showy, flamboyant new leaves, and in winter with its berries, which are now a whitish-green colour; and to stand under one and look up at the canopy strung with small white beads is magical.

It is breeding season for the black kites and a pair has started building a nest on the very top of a huge neem near our house. From a distance the nest doesn't seem to be very sturdy, being a simple jumble of twigs, with some pieces of paper stuck randomly, but there is a definite method to the layout, and the structure is able to take the weight of the birds as well as withstand the occasional gust of wind. These kites—subspecies *govinda*—are year-round residents, though I find that they are more visible in the winter.

17th

The morning was again dark and overcast and cold, and I had to switch on the lights in my room, but by afternoon the fog lifted and the sun came out. Bright blue sky, with feather clouds and eagles wheeling far overhead; a brisk wind pushed the clouds along gently so that one pattern followed another; a long feather, maybe that of an eagle, followed by a shorter bulbul-sized one. These are cirrus clouds, ethereal looking, and form high up in the troposphere.

The dahlias are doing beautifully. Randomly selected baby plants have grown and blossomed into different varieties; open-petalled shaggy ones, more compact red and white beauties and, my favourite of all, light peach-coloured blooms with petals like penne pasta. No exaggeration here, individual petals are the exact shape and size of a penne!

18th

The ylang ylang shrub/vine that I planted a few years back has grown to more than six feet tall and bears clusters of the most unusual five-petalled greenish-yellow flowers all through the summer. The blooms are highly fragrant and are widely used in the perfume industry. The small glossy green fruits begin to form in autumn and turn lemon yellow this month, falling almost instantly on ripening. They look exactly like miniature Sorrento lemons, but the surprising thing is that I have never seen a bird eating one.

The ylang ylang shrub, also called manorangini or Hari Champa, is found all over India and Southeast Asia.

20th

At the corner from the goolar—the big one with a family of owls—where a side road leads to the gate, stand two trees, a neem and a goolar, growing from the same spot, and the contrast between them, especially at this time of the year, is striking; the neem with its greenish-yellow leaves and dark bark, and the goolar with its dark green leaves and whitish bark. Ebony and ivory!

22nd

Went to a lovely nursery in Gurgaon, or rather, a row of small nurseries located in a quadrangle, with one side housing tin sheds chock-a-block with ceramic pots and houseplants, and the rest of the area swathed in tiers of flowers. Beautiful pansies, violas in colours I have never seen before, golden gazanias with sunlight trapped in their petals, cinerarias—in all the shades of blue; azure, cobalt, cerulean, indigo—fragrant stocks, primulas in rainbow hues, ranunculus the size of roses, dahlias growing with riotous abandon, phlox tumbling out of their hanging baskets and nasturtiums in an almost gaudy shade of orange. A gay prismatic carpet in an otherwise monochromatic urban jungle overgrown with grey high-rises and impersonal facades.

I bought pots of herbs; thyme, basil, rosemary, parsley, celery; and chillies that change colour from green to orange to purple.

A very interesting addition to the garden centres this year are calla lilies in different colours; pink, cream and a dark blackish-purple. The gardener assured me that they would be able to survive the summer if shifted to the shade, so I indulged myself and bought all three.

23rd

Bright and sunny with a crisp biting wind, which kept up a brouhaha in the peepal all night long, before pushing frosty pink clouds across an icy blue sky, till the orange of daybreak diffused gradually into a transparent luminosity. The leaves of the peepal are natural wind chimes, responsive even to the mildest breeze, and I sometimes think I should string up some to hang on the balcony.

We went to the club for lunch to make the best of the sunny afternoon. One of the pleasures of Delhi winters is being able to eat outdoors, under a blue sky, with eagles flying far overhead—or not so far overhead. Black kites and house crows, bold and ever opportunist, swoop low over the tables laden with fish and chips and chhole bhature, their sharp eyes taking in even the smallest unattended titbits. Malcolm MacDonald, in his book *Birds in My Indian Garden*, gives an amusing account of one such encounter in his garden in Delhi in the month of January[6]:

> They [the Kites] are superb marksmen. One afternoon I sat alone on my lawn sipping tea and munching delicious cake. As I held a slice between a finger and thumb, and was about to pop it into my mouth, my attention was distracted by two mongooses playing hide-and-seek in a near-by flower-bed. Turning in their direction, I momentarily left

my hand upheld in mid-air, clasping the cake. Suddenly I felt the food gently extricate itself from my grasp, and a fraction of a moment later my hand was empty. I could no longer feel the cake between my finger and thumb. Then I saw it about twenty yards away, flying rapidly from me in the claws of a retreating kite! Every now and then the bird hesitated in its flight, bent its head, ripped a morsel of cake away in its beak and swallowed it—evidently relishing it as much as I had intended to do. The robber had swooped down from on high, glided above my fist and, as it passed, neatly picked the delicacy from my grip without so much as brushing my thumb or finger or any other part of my hand. It was an impressive demonstration of a kite's sureness of vision, accuracy of aim, precision of movement and delicacy of touch.

We had a similar experience the other day, whilst out for lunch. As the waiter was carrying our paneer tikkas to us, a kite swooped and with surprising nimbleness for such a large bird, made away with one piece. Such dexterousness—I'm tempted to call it sleight of hand!

25th

Took a drive outside Delhi on a cold, foggy, grey morning. Fields of marigolds streaked across the green, unripe wheat like rivers of gold, their brightness alleviating the gloom somewhat. The ubiquitous genda, so completely synonymous with India and Indianness. It is everywhere; heaped in baskets outside temples, garlands strung around idols and hung from houses during celebrations, and even offered to welcome chief guests on stage. No Hindu auspicious occasion is complete without the presence of this bright flower. Golden curtains of marigolds decorate homes during marriages and are hung outside doors as torans, often interspersed with strings of mango leaves, which are then left to hang sometimes for months on end, until it is time for the next festival. I have seen a solitary genda lying at the feet of tiny idols kept in equally tiny niches in most shops, the shopkeeper using it to sprinkle water on the gods while reciting a hurried mantra! It is surprising, then, to realize that the flower was introduced to India by the Portuguese. It has had a long association with Christianity and is believed to have got its name from 'Mary's Gold', the flower being offered instead of money at Mary's altar.[7]

The colours of the marigold find deep resonance within Hinduism; orange signifies renunciation and also courage and sacrifice, while yellow symbolizes the sun and purity. These days, one can buy pots of marigolds in various hues and sizes and for the past couple of years, I have been bringing home a beautiful burgundy coloured variant. But at the nursery I visited earlier in the month, I succumbed to the temptation of getting plants with deep rust and orange flowers, the petals arranged in an almost ombré formation. Surprisingly, however, I have never seen marigolds sold at florists or indeed sold with their stalks as other flowers are, which is strange, because the stalks are strong and the flowers comparatively lived long. Clearly this is a flower meant for higher things!

28th

Full moon night! At 8 o'clock the moon is low on the horizon and has a distinctly red tinge, probably because of the slight haze in the atmosphere, but within a couple of hours, it rides high in the sky, and shines brilliantly, almost like the overhead sun at noon. Imagine the full moon shining this way for 24 hours north of the Arctic Circle! This is the wolf moon of the west, probably so named because it was thought that the cold made the wolves howl. For the Hindus this is the Pausha Purnima, a very auspicious day when thousands take a dip in the Ganga and Yamuna rivers.

My love affair with the moon is an enduring one; I find it endlessly fascinating and sometimes, in my fancy, imagine that its silver glitter casts as much of a spell on me as it does on the tides.

30th

Although the magpie-robin has kept up its call all winter, I heard one sing after some months. This could mean only one thing; spring is on its way! Yesterday we had more than a dozen red-whiskered bulbuls on the feeder table again, winging between food and water, serenading us with their mellifluous song all the while.

Cold, sunny, very windy; my favourite kind of weather. It has been a very cold and wet fortnight and it is nice to have some warmth again. Most trees are looking straggly and tired these days, with yellowing leaves and bare branches. Two kaniar trees near our house are in flower though, the purplish-pink blooms glowing like beacons in the gloom.

February

From early February onwards the sunlight grows gradually warmer, which has its effect on certain species of birds. Spring is in the air, and they respond to its suggestive caress. Indeed, some of them begin to feel like imparting caresses themselves.

Their mood is revealed in both sights and sounds. By February several of them have donned their courtship dresses. […] These fresh sights are accompanied by appropriate sounds. Phrases of love begin to echo through the garden.[1]

1st

A very wishy-washy start to the month that is the herald of spring. The average temperature for February hovers between a very pleasant 11°C and 24°C, while the average humidity settles at about 50 per cent. This is a time for balmy days and clear blue skies; a time for love, and our avian friends often ditch the feeders and the mundane act of eating to pursue more important matters of the heart; a time for flowers and birdsong and coffee at open-air cafés before the days become too warm.

But to come back to today; a weak, barely-there sun coupled with a brisk wind makes the afternoon rather cold, and I pull my shawl a little bit closer. But the barbets are calling in the saptaparni and that can mean only one thing—spring!

2nd

February started with us travelling to Chandigarh for a family wedding. Getting out of Delhi is always a problem, but once the traffic snarls, roadblocks and stray cattle have been tackled, it is a pleasure to be driving under the open sky. Unsurprisingly, a lot of construction has taken place on both sides of the highway over the last several years, and every time we hit the road, the distance between towns and villages seems to have shrunk some more. And I fear that one day, in the not-so-distant future, the open road will become just another concrete strip through a concrete jungle.

I remember the time as a child when the drive from Chandigarh to Delhi would be a huge adventure, a never-ending picnic, complete with sandwiches and cake and the occasional rationed sip of coffee; when fields of green stretched away as far as the eye could see, and the landscape was dotted with 'kos minars', which lined the road on both sides. Kos minars are relics of the Mughal period and basically served as milestones. They were erected every few kilometres (a kos measures roughly 3 km) on GT Road. Akbar issued an order in 1575 for building a minar at every kos right from Agra to Ajmer along this road.[2] A kos minar is a solid round pillar that stands on a low brick platform and is dotted with tiny windows, which, I remember my mother telling me, gave out light from the diyas lit at sunset. A sort of beacon or lighthouse on land! Kos minars served as route markers for travellers, who could be sure that they were on the correct path if they crossed one at regular intervals. Some historians are of the opinion that serais or travellers' inns were also located nearby. Horses and riders were stationed at many of these minars to serve as 'dak' runners, thus providing an efficient means of delivering royal messages.

Sadly, most of them have now vanished, demolished in the drive for commercialization; for widening the roads and constructing dhabas every few kilometres. The road is lined for the most part, at least where it has not been taken over by dhabas, by eucalyptus trees, which have grown to a considerable height over the years. Tall and slender, constantly swaying to the tune of the breeze, they have entranced me since I was a child. Their woody fruit was just the right size to become a goblet for a favourite doll! There were different types to choose from; goblets with straight or serrated edges, which made the latter look like small flowers. And the flowers themselves wore small pixie caps! The leaves of some varieties have the delicious fragrance of lemons, while those of the others, I have always associated with the common cold and blocked noses. However, I confess that I find the different varieties of eucalyptus difficult to tell apart by the mere colour of their bark and the shape of their leaves. They all looked the same to me from a moving car, but what I did notice was that most of them were in flower; cream-coloured pompoms against a bright blue sky.

Since the eucalyptus is a hardy tree and can survive in difficult conditions, it thrives in the low-lying, often waterlogged, narrow stretch of land between the road and the fields beyond. Water hyacinth grows abundantly in isolated pockets, and I saw a couple of water rails and herons wading in the shallows, foraging for food. Most of the fields were sown with wheat and sparkled emerald green in the sunlight, the colour a perfect foil for the occasional late-season sarson still standing. Fields of green and gold!

4th

It is noon and the avian world has been all aflutter for nearly an hour now. Pandemonium breaks out in one corner of the neighbourhood, with the babblers shrieking and the crows cawing and the squirrels setting up alarm calls, before shifting to another corner after a momentary lull. A common buzzard, a winter migrant from the Himalayas, has decided to make our area his hunting ground this morning, perching for considerable time on the saptaparni outside our house, sending a wave of terror pulsing through the garden.

I have been observing for some time that large shadows, like that of an aeroplane flying overhead, frighten the birds and squirrels into scampering away instantly. Maybe it has to do with an evolutionary instinct that tells them to fear shadows that could mean big birds like raptors. Our house is near the airport, so low-flying aircraft—taking off or landing—cast a momentary but distinct shadow on the ground. This is especially noticeable during those winter days when dense fog delays flights till mid-morning so that when the fog clears and the sun shines through, the aircraft take off one after the other in quick succession.

5th

Basant Panchami! This festival is celebrated on the fifth day of the waxing moon in the Hindu month of Magha. It is a sort of prelude to spring; the fields are golden with sarson, the days are balmy enough for flying kites, gardens are full of flowers and halwai shops are laden with mithai. As a child, I looked forward to this festival because of the sweet saffron rice, full of plump raisins, that my grandmother made every year.

However, as Basant Panchami is very cold and foggy this year and gives no inkling of the coming season, I might as well stay home and try and copy my gran's recipe!

6th

Who would have thought that the ordinary, modest-looking shahtoot or mulberry tree, standing a few houses down our lane, could transform itself into something so pretty, especially after having shed all its leaves last month? It has been cloudy since last evening and even rained intermittently through the night. I love hearing the pitter-patter of the raindrops on the windowsill, more so at night when everything is so quiet. This morning around eleven, when I went for a short stroll in the neighbourhood to check out the clouds gathering on the horizon again, I noticed the shahtoot. The branches are laden with tiny green leaves, and in the special diffused light of a cloudy, rapidly darkening mid-morning, they look gorgeous. The lime green of new shoots, backlit against the emerald of a tree growing just behind this one, transformed our ordinary toot into something almost magical.

14th

For the lucky ones who live in a haze-free atmosphere and also have a telescope as a bonus, seven planets are visible in the night sky these days. On an overnight trip to a more rural location, at about 5 o'clock in the morning, before sunrise, I went out to check the still-dark sky and was rewarded handsomely for my effort. A most spectacular Venus shone to the east, and I can safely say that it was the brightest celestial object I have ever witnessed, brighter even than most full moons, for this was a different kind of light, concentrated and bedazzling. As I stood there, awestruck at the thought of the distance this light had travelled to reach me, I got lucky a second time! Also in the east rose the planet Mercury, equally beautiful in its own way, with a distinctly yellowish hue as against the silvery light of Venus.

15th

What an entertaining morning it has been! It is not often that you get to witness a pitched battle between three pairs of mynas, complete with screams, screeches, aerial sallies and vicious pecking. The handsome goolar tree just outside our house is beginning to effloresce and the young leaves are enchanting; a beautiful lime green when backlit. But today it was not the leaves that held my attention; it was the raucous shouting and scolding, tumbling and tearing that kept me absorbed for well over half an hour.

Halfway up a central branch of the tree is a biggish hole that has, for the past few weeks, been inspected by parakeets and mynas alike. Today morning, while having my cup of tea, I noticed a pair of rose-ringed parakeets examining the prospective nest carefully; the first one entered, stayed some time and then the second one went in and pottered around. But eventually, they found it not up to their expectations and left. Shortly afterwards, a pair of mynas subjected it to their scrutiny, and decided that it was the perfect place to bring up juniors. That's when the trouble started. Another pair of mynas followed soon, and then a third, all with the same idea. A violent battle ensued, which I am sad to report, claimed its first casualty after about 20 minutes of the start of proceedings. The cleaning boy, who was sweeping underneath the tree and who had long since stopped all activity in order to get a bird's eye view of the feud, recoiled in shock when the martyred bird landed at his feet. Picking up the poor thing, he put it in his wheelbarrow and tried to broker peace by waving and shooing at the rest. Silence reigned for some time and all seemed well. But after 10 minutes or so, I again heard the familiar battle cry. A pair of mynas, after re-strategizing, had launched an aerial strike on the enemy, and this time the birds tumbled and fell to the ground fighting, wings entangled and beaks sharpened. They remained on the ground for a full five minutes—oblivious to the boy standing nearby—alternately brawling and sitting still. The lad seemed bemused at first but finally gathered himself together and threatened them with his broom, and that's when the ceasefire came into force. But not for long, I'm sure; mynas can be terribly quarrelsome when the mood takes them, and with their nesting season lasting from February to December, one can be sure of nearly year-long entertainment!

16th

Very windy but sunny; in the late afternoon, I spent an enjoyable half hour watching the wind chase clouds across the beautiful blue sky. High-level cirrus pulled into commas and apostrophes and fantastically-shaped feathers!

> 'O, it is pleasant, with a heart at ease,
> Just after sunset, or by moonlight skies,
> To make the shifting clouds be what you please.'
>
> —SAMUEL TAYLOR COLERIDGE, *Fancy in Nubibus*

The silverbills are out in full force, and I counted nearly two dozen at the feeders yesterday. They are energetic little birds and are surprisingly bold for their size, willing to join in with the doves and parakeets. They have a strong 'follow thy leader policy'; several of them will crowd on a branch, keeping up their incessant chirping, all the while waiting for the pioneer from amongst them to land on one of the bottle feeders. Immediately therein, all of them will try to crowd on the same spot, jostling and pushing to get to the grain. Twice last month, I was pleasantly surprised to see a pair of scaly-breasted munias feeding with a flock of silverbills in a corner of the garden, where I keep a biggish terracotta pot filled with mixed grains. Both times it was their behaviour that set them apart; disturbed by some small thing, the silverbills all flew up together, whirling overhead, but two birds remained on the ground, feeding nonchalantly. A closer look revealed the distinctive markings on the chest and the lack of white on the rump.

20th

A timeless scene. New leaves on the goolar, glistening in the golden haze of the late afternoon sun, parakeets chattering loudly, the raucous call of the treepie, our resident spotted owlet sunning himself on his favourite tree stump across the road, a squirrel on the peepal having spotted a cat nearby, giving out a series of loud warning calls. A black-rumped flameback woodpecker just metres from where I'm sitting, foraging for dinner along the trunk of the same peepal, is unconcerned about the alarm the squirrel is raising. In the golden hour before sunset, its colours are all the more vivid, the yellow back and the red crest saturated with deep, rich pigments.

> I find these woodpeckers fascinating. They look so ornate and outlandish, like birds out of a fairy tale; and as they run easily up the sheer surface, or slip down it, with no change in their rigidly held pose but for quick, sideways transportation, they do not look like birds at all. Their movements have that quality of change of place, without obvious, free use of limbs that suggests clockwork.[3]

I love these early evenings when the sun has not yet set, but the light has turned a luminous mellow golden and shines now only on the treetops. I find that a light breeze often picks up around the same time, and on a day like today, which was overcast and stuffy, it feels delightfully cool—the perfect spring zephyr!

The garden is glorious these days, the petunias are overflowing their hanging baskets while the pansies, in their purple and golden robes, spill out of their own pots onto their neighbours so that it is difficult to tell where one ends and the other begins. We bought two tiger lily plants, early shoots really, in pretty terracotta containers, hoping that at

least one of them would turn out to be orange, but they are both white. It is a good thing in a way because they make the twilight sparkle, but my daughter, who was insisting all along that they were of different colours, is disappointed. The magenta phlox is resplendent in its blue basket, but the star of the show is the single pot of nasturtium, which is winning all competition hands down. The bright orange flowers have somehow managed to climb up and hang down from the basket at the same time, and are a magnet for the bees that hover around all day. The children, however, are very wary of the bees, and rush in the moment one appears. The daisies are doing well too, although the ones that were planted earlier seem to be past their prime now.

> *'I must have flowers, always, and always.'*
>
> —CLAUDE MONET

21st

A brisk west wind, making the late afternoon seem colder than it actually is. This is the wind that signals, subtly, the change of seasons, from winter to the brief spring and then onto the seemingly endless summer. It won't be long before the neon-green leaves on the goolar, now appearing to be little more than tiny pinpoints of light, transform themselves into a lush canopy.

The birds have spring fever; barbets, who are the official herald of warmer weather, have been unusually vocal for the past couple of days, a courting pair of koels is very noisy in the saptaparni and the parakeets have begun their annual battle for nests, screeching and squabbling the entire time. On the leafless gulmohur outside our house, a pair of parakeets are rubbing beak to beak, with the male, who is perched slightly higher than the female, extending his foot as if to reach out and touch his mate. They are engrossed in each other, oblivious to their homeward-bound friends and relatives flying overhead in the early evening.

> 'And then within these mountain-forest reaches,
> Skilled to distract saints' thoughts from heaven above
> The young awakening Spring now yawns and stretches,
> Belov'd companion of the God of love.
> On every blossom arrow he created,
> Feathered with leaves and tipped with mango-flame,
> The fletcher Spring the owner designated,
> Writing with bees the God of love's own name.'
>
> —KALIDASA, *Kumarasambhavam*

24th

The seasons are changing and the signs are everywhere; the air has a distinctly warmer feel, leaves are drifting down from the neem in a golden flurry, mango trees are heavy with baur, the kosam is clothed in bright scarlet and the semal (silk cotton tree) is in bloom.

'Sprays of full blown mango blossoms—his sharp arrows,
honey-bees in rows—the humming bowstring;
Warrior-Spring set to break the hearts
of Love's devotees, is now approaching, my love.'

—KALIDASA, *Rtusamharam, Canto VI: Spring*

A flowering semal is a beautiful sight, the branches smothered in red flowers, striking against the cloudless blue skies that are so typical at this time of year. The tree, which towers over most others in its neighbourhood, is an arresting one, and the thin pleat-like flanges at the base of its trunk make it unmistakable. It becomes conspicuous by January when the leaves thin out and plump buds start forming on the branches. The flowers, when they appear in February, are a lovely sight and range in colour from bright red to coral to different shades of yellow. A tree near our house has bright gold ones, and is my favourite, while two others nearby are a lighter, sherbet-lemon colour. Another beautiful tree to enjoy this month is the Hong Kong orchid tree, the most beautiful bauhinia of Delhi. The large magenta flowers are unmistakable, although I confess to having trouble in identifying this tree from its other two relatives, the kachnar and kaniar, when it is not in flower.

25th

The babblers have been in for a treat today. Yesterday evening, I noticed that the wooden stakes, which were put up to support the sweet peas, had termites on them. Huge columns of termites had come up overnight, and while I was worrying and pondering over the problem because I don't like using chemical pesticides, I realized that I needn't bother myself at all. A babbler discovered the termites, and almost immediately a whole army of them descended on the sweet pea bed and demolished the enemy. It was an amusing sight; the birds working their way down the stakes before turning their attention to the ground for any leftover treats. An eco-friendly solution to a pesty problem!

26th

The wind is frisky this evening and plays with the leaves of the neem, hurling them high, and holding them suspended mid-air for some time, before letting them drift down slowly, in stages, swirling and eddying along the way. There is, however, always the possibility of a sudden gust leading them astray again!

An open window, especially one that has views of the sky and the clouds, and is fortunate enough to have a tree positioned right outside, is a source of endless pleasure, as I am rediscovering for the umpteenth time this evening. It has been a bright, haze-free day, as windy days usually are, and even though the sun is well above the horizon and dusk has not yet set in, I can clearly see the moon waiting in the wings!

27th

An absolutely gorgeous peach-hued full moon, the last full moon of winter, rides high and dominates the cloudless night sky with its size and luminance, seeming impossibly near, its halo bright and clear. For us in the northern hemisphere, the full moon this month stays out longer than 12 hours because it follows the high path of the summer sun.

The February full moon is traditionally called the snow moon by the Native Americans because this is the time of year when it snows the most in temperate latitudes. In Europe this moon goes by the name of storm moon as well. For the Hindus this is the Magha Purnima, a day of special significance, which marks the last day of the month of Magha. According to both the Hindu and Buddhist calendars, this is the month when the full moon is in the constellation that contains the star Regulus. Magha Purnima is marked by ritual bathing at the confluence of three rivers, or 'triveni sangam'. Every 12 years the Kumbh Mela is held at one of these locations and every year Magha melas are held at all such confluences in North India. For the Buddhists also, this is a very important festival, and they celebrate it with Magha Puja, this being the day when the Buddha gathered his followers to create a community or sangha.

28th

A rainy winter afternoon, tucked in under a chunky hand-knitted throw on my window seat, a mug of frothy mocha loaded with swirls of dark chocolate, and I find my thoughts rambling. I ruminate over certain words and their almost magical ability to transform the setting, much like the special effects used on the stage during a play, changing it from rainforest to tundra in a moment. Like *komorebi*, which, in Japanese, means the sunlight that filters through the leaves of trees, and which instantly transforms the rain-drenched cobblestone path outside my window into a sun-dappled forest floor. Unfamiliar words, alien even, but possessing the uncanny ability to say exactly what I want to or to perfectly describe the mood at a particular moment. So right now, I'm a *nefelibata* (Spanish), living in the clouds of my own imagination, and my thoughts that wander through these clouds are *nubivagant* (Latin). Does that give me a nimbus, an aura of my own? And the *rim jhim* (Hindi) outside transforms into a spindrift on the windowpane, swept up by a wild sea, and I finally have a name for the sound of wind in the treetops—*psithurism* (Greek)! The brown-headed barbets, who have set up a very energetic duet since the morning, are in the throes of *forelsket* (Norwegian), the elation felt as you first start falling in love!

As the day turns to dusk and it grows darker outside, I want to count the stars in the night sky and trace a *mangata*

(Swedish), the road-like reflection of the moon in the water, on the ocean of my imagination. To put my arm out and measure a *gurfa* (Arabic), the amount of water I can hold in one hand. And as the dark, louring clouds settle lower, and old memories surface unbidden, I am filled with a sudden *hasrat* (Urdu), an intense desire for things that are forever lost to me.

Words are the chroniclers of our times, and as we change, so do they. They make me a procaffinator these days, and I find myself unwilling to do even the simplest of tasks till I have had my morning cup of tea. I may even need a *tretar* (Swedish), a three-fill, to function efficiently. They also make me judgemental, and I have developed the deplorable habit of looking at people who are not into embroidery or sewing with 'craftasm', which is basically self-righteous, sarcastic disapproval!

All in all, I have spent a pleasant evening indulging in some *boketto* (Japanese), gazing vacantly at the rain and the trees in the distance, without really thinking about anything in particular.

> *'I am beginning to learn that it is the sweet, simple things of life which are the real ones after all.'*
>
> —LAURA INGALLS WILDER

March

'The trees began to whisper, and the wind began to roll,
And in the wild March-morning I heard them call my soul.'

—ALFRED LORD TENNYSON, *The May Queen*

1st

I always think of March as being the first of the hot months, a period of transition from spring to scorching summer. The average temperature hovers between 31°C and 17°C. Mornings and evenings are usually pleasant enough to sit outside, but even then, I find that I need a table fan from mid-month onwards. March sees a considerable cloud cover, and the percentage of time that the sky is overcast is nearly 25. However, it is not a rainy month and the average rainfall is barely 24 mm. The length of day is increasing throughout the month, at the rate of nearly 1 minute and 40 seconds. The wind direction is mostly from the west.

There is a frenzy of growth in the flower pots and beds, with the stragglers pushing through the soil in a rush to bloom and complete their life cycle before the increasing heat gets to them. This impulse is primordial.

> 'Because it would, because it must,
> Because of life and time and lust.'
>
> —WILLIAM ERNEST HENLEY, *Hawthorn and Lavender*

There is equal excitement in the avian world, with our local birds having found partners. Berries are ripening on the fig trees, providing the nesting pairs with an ample food supply. This is also the month of *patjhad*, when most deciduous trees shed their leaves—neem, pilkhan, harshingar, amaltas, peepal, palash and so many more—painting the roads in shades of ochre and burnt sienna. This reverse timing of autumn colour as compared to the Western world is nature's way of coping with the coming dry hot summer. Paradoxically, however, this is also a time for renewal and regrowth. Some fig trees, including a peepal a couple of houses down our lane, are already in new leaf, while others are still getting rid of the old yellow ones. An avenue of pilkhans that I cross every day is like a river of molten gold and copper, while the bright red of new kosam leaves shines like a beacon against the still bare branches. My plumerias or temple flowers, which had withered beyond recognition during the winter, are sprouting new growth. I am especially delighted with the red champa, which I had almost given up on, and which has finally pushed up a tiny green shoot. A month of endings and new beginnings!

3rd

It rained intermittently the whole of last night, showers that were soft enough to not keep you awake, but strong enough to not let you sleep. When I woke up in the morning, it was still drizzling—in a gentle *poohar* sort of way—but even before I had finished my tea, it stopped and the birds came out to feast on the insects and worms that the rain had flushed out from their homes. The birds appeared in droves; magpie-robins and Indian robins, red-vented and red-whiskered bulbuls, babblers and parakeets, treepies and barbets. Half an hour back, a young girl walked past, carrying provisions in a bag that had clearly seen better days. What was astonishing was the birds following her in a Pied Piper sort of way. This demanded an immediate investigation and at the risk of seeming impolite, I followed the trail and was amused to see chana dal spilling out in a thin trickle from a corner of the bag. Even torn bags have their uses!

Our resident family of three spotted owlets shifted to the side of the goolar directly facing us last month when a storm broke the branch they had their home on. The new quarters are housed in a hollowed-out stump midway up the tree, well camouflaged by fresh leaves. But one of the owls is a rebellious sort of fellow, not following the dictates of his tribe. The whole day, instead of sleeping in his hollow, he suns himself on a thick branch of an old neem on the other side of the road, which incidentally also has a hollow where parakeets nested last year, and where he sometimes takes up temporary lodging, keeping a watchful eye on the comings and goings in his neighbourhood. He reminds me of my son, who also likes to retire to his room when he has had enough of us, except that the little owlet flies in a funny, bobbing way, as though he is riding a wave, while my boy stomps off disdainfully!

4th

The early evening is glorious; blue skies, spring zephyr and, most of all, that special quality of light, which is for me the most defining factor in a day. Like today; mellow, golden, eternal light, which magically transforms the new leaves on the mulberry into a luminous lime green, and I am reminded of these lines that I love:

> 'The sun from the western horizon
> Like a magician extended his golden wand o'er the landscape;
> Twinkling vapours arose; and sky and water and forest
> Seemed all on fire at the touch, and melted and mingled together.'

—HENRY WADSWORTH LONGFELLOW, *Evangeline*

One of the neem trees growing across the road has died, killed no doubt when the tube well—or storage tank or whatever—was being constructed and the entire area was dug up. I would have thought that a fully grown neem would have been resilient, but the construction team proved too much for the poor thing. I mourned the loss of my tree, my haven of greenery and birdsong and shield against prying eyes from across the frontier! But I have gradually come to realize that nothing in nature is without its purpose. A pair of parakeets has started digging for a nesting hole in one of the side branches while a couple of treepies have become regular visitors, probably feeding on termites and other insects that inhabit dead wood. Bulbuls use the bare branches as a lookout perch, while mynas cavort and create their usual cacophony.

A mulberry nearby is clothed in new leaves. The fruit has grown long enough to be clearly visible and attracts birds the whole day. My red hibiscus is flowering profusely and is a magnet for purple sunbirds, while butterflies prefer the pansies and violas and the pinks.

6th

A bright sunny spring day, and everything shines with technicolour brilliance. The sky is a beautiful shade of blue, which lightens in ombré waves towards the horizon, creating the perfect backdrop for the yellowing leaves of the neem. The rose-ringed and alexandrine parakeets are full of energy, screeching and squawking as they fly about busily, rather incongruous against the motionless, almost torporific yellow-footed green pigeons sitting quietly on a branch of the nearby Persian lilac, making a pretty picture among the lilac flowers.

7th

Last week, while walking in the park, I heard the distinctive tap-tap of a woodpecker. Expecting to see the black-rumped flameback, I was instead treated to the sight of a brown-headed barbet excavating a nest hole. It was hard at work, tapping and hammering, chipping and chiselling, with tiny shards of wood flying in all directions. Its mate (which I am pretty sure was the male) was sitting on a branch immediately to the left, probably providing moral support. So engrossed was the bird in its work that it paid no heed to inquisitive passers-by like me, who inched as close as it was possible without getting some timber into their eyes. However, every few minutes it would draw away from the tree slightly and look around, perhaps in a bid to alleviate the boredom of constant hammering, and it was then that its breeding plumage of bright orange circumorbital skin and reddish-orange beak would be visible. It was a bird in a hurry, with the nesting season—which lasts from February to September—upon it and the nest not even half done. The small hole it had excavated would have to be widened and deepened into a tunnel many inches deep, at the bottom of which the eggs would be laid.

8th

A beautiful late afternoon, sunny with a cool breeze, so typical of this time of year. 4 o'clock onwards, when the sun has withdrawn from our little corner, taking with it the last of its disagreeable midday heat, the garden suddenly becomes very pleasant and I spend a delightful hour admiring the flowers in all their riotous glory and watching the birds at the feeders. New leaves are starting to appear on the Persian lilac and together with the nut-coloured berries still hanging in bunches from the topmost bare branches, they form a very pretty picture against the deep blue sky. The late afternoon sun adds high gloss to the vignette, but try as I might, I am never able to capture that fleeting moment of beauty on camera, making me ponder whether we assimilated wondrous experiences better before we had the luxury of instantly capturing them on a device!

A very peculiar sighting in the garden today. Accompanying the flock of almost two dozen silverbills was a reddish bird, with dark scaling on its underparts. My first thought was that it was a scaly-breasted munia, pairs of which occasionally do visit the garden. But this bird had a pinkish wash all over and was also tailless! After much consultation with fellow birders, the tentative conclusion reached was that it was indeed a scaly-breasted munia, which some unscrupulous bird trader had tried to colour, both to give it an exotic appearance and to circumvent the bird pet trade laws. It was probably an escaped pet, a theory given credence by the fact that it was a bold little creature, continuing to forage on the ground even after the silverbills had departed.

9th

Around 5 o'clock every evening, as the sun prepares to set, I notice the mellowness of the season all the more. The sunlight is gorgeous at that hour, streaming in through the windows, settling itself like a blanket on my bed, pushing away the last vestiges of darkness from the room. While Delhi, like the rest of India, does not have the traditional spring of the West, late February till the middle of March is a beautiful period. Winter is over (although I do love the cold, grey winter days), flowers are in full bloom and the trees are clothing themselves in new finery.

> *'It's spring time.*
> *A fragrant breeze blows. Fresh, tender leaves*
> *Sprout on the branches of the trees.*
> *The koel sings a sweet and passionate song.'*
>
> —BHARTRIHARI, Atha Sringarashatkam

One tree that grabs my attention these days is the dhak or flame of the forest. The compound leaves with three large leaflets have given rise to the saying *dhak ke teen pat*, which is somewhat similar in meaning to 'a leopard never changes its spots'. The leaves may not change the number of their leaflets, but they certainly give way to gorgeous flame-orange flowers in March. My one regret is that the flowers grow high up, near the top of the tree, and the only way you can see one closely is if you pick up a fallen specimen. I love the velvety softness of the inner side of the petals and the way it contrasts with the hairy outside. Sometimes one gets to see a tree with light orange flowers, or if you are particularly lucky, with beautiful golden yellow ones.

> The timing of the Holi festival must have something to do with the flowering of the dhak, for the popular yellow dye they produced was synonymous with the spring festivity and often used to produce colour for the occasion.[1]

The more I see trees, the more I appreciate that there really is a time for everything under the sun; there is a time for lying low, covered only in drab brown foliage, and there is a time for being brilliant, for letting your light shine through.

10th

The late afternoon light is very forgiving, gilding every leaf, softening every angle. It bounces off the pansies in their pretty pots, lengthens the shadows of the ferns in their hanging baskets and makes the air iridescent. It is a paradoxical light, sharp and mellow at the same time, as if nature has reserved the best for the last.

There is much activity in the avian world, a sort of pre-dusk chorus. The coppersmith barbets are singing their last duet for the day, the magpie-robin has retired to a branch in the goolar from where it occasionally trills a few short, sweet notes, parakeets screech overhead as they fly home to their roost and a pair of tailorbirds have dived into the bougainvillea. A short while ago, I could hear the mellifluous call of the koel—who has become active once again—from the neem some distance away, and the more guttural tone of the treepie from the peepal next door, but as the sky softens and gold gives way to the grey of dusk, birdsong gradually falls silent, and taking a cue from nature, I make my way indoors too.

11th

An impromptu early morning visit to Chandu Budhera near Gurgaon proved to be very rewarding. The migrants are still there and we saw flocks of bar-headed and greylag geese, stints, ducks, coots and sandpipers. But the highlight of the visit was the sighting of a spotted crake, an uncommon winter visitor to Delhi. We got lucky; it was foraging in a dense cluster of reeds in a marshy stretch along the road.

It was a beautiful day; the wheat was golden in the fields, and the sky was blue, with swallows skimming in the liminal zone between the two. Skylarks were on the wing, and we saw four different kinds; crested lark, Oriental lark, ashy-crowned sparrow-lark and Bengal bush lark, all of which are year-round residents of Delhi. This was my first sighting of the diminutive ashy-crowned sparrow-lark and I was once again awestruck by the perfection in nature: the bird was foraging in a muddy kaccha track right in front of us and was so beautifully camouflaged that it took me a minute to spot it. Black-eared kites were there by the dozens; a subspecies of the black kite, they are residents of temperate regions, migrating to the tropics during the winter. These are large birds, with a wingspan of 140–155 cm, and fly to Delhi from their breeding grounds near the Altai mountain range. A recent study, in which 14 birds were tagged, has shown that the kites use the Central Asian Flyway, travelling over nine countries to reach their destination.[2] What is remarkable is that they cross over the Karakoram Range at an altitude of over 25,000 feet, with many individuals flying past K2, the second-highest mountain in the world.

For the last couple of years, I have been noticing that two trees appear out of nowhere in March, both a 20-minute walk from where we live. They are, of course, right there the rest of the year too, but veiled in obscurity. This month, however, there's no ignoring them; smothered in pale pink blooms, the quickstick trees shout 'Look at me', and indeed they are a delight for the eyes. The baby pink flowers growing on the end of the bare branches cover the canopy in an ethereal haze. Pradip Krishen's excellent book on the trees of Delhi informs me that they are widely cultivated in the tropics to provide shade in cocoa plantations.[3]

12th

One of my favourite things about March is the shedding of the neem leaves. I love the way they cascade like a shower, raining down at the slightest touch of the wind. Or maybe raining is not the correct word, because rain is hard and usually involves very little drift and little or no play. But neem leaves are frisky, riding the wind, drifting down slowly, almost like snowflakes, gilding every available surface a lovely copperish gold.

But not everyone shares my love for them; certainly not Guddu, who has the duty of sweeping the road near our house. Yesterday, as I was lazily enjoying a mid-morning latte on the day bed next to the window, I saw him dutifully sweep up the carpet of dried leaves into a heap on the curb side, from where he picks them up later in his wheelbarrow. But it was a windy day, and no sooner had he cleared a patch that a new set of leaves would dance down, covering it in gold once again. After a few rounds of this battle for supremacy, which the leaves were winning hands down, Guddu began to get exasperated. Finally, stopping all work, he leaned on his broom and looked up at the canopy in a bemused sort of way, as if counting the number of leaves still left on the branches. Realizing that the odds were stacked heavily against him, he withdrew gracefully to the edge of the battle zone, and making himself comfortable on a small pile of bricks, lit a bidi, sending all his cares up in smoke!

13th

The bakain or Persian lilac tree, growing just outside our gate, is in bloom and is a delight for the senses. It started shedding its leaves in January and for that entire month, the driveway was covered in golden drift. The flowers are gorgeous, highly fragrant and a delicate lilac-white. They hang in clusters and are very similar to the flowers of the harshingar, except for the colour; the central column of the former being a deep purple instead of the familiar orange of the latter.

At first glance, the Persian lilac is easily confused with the neem; the leaves, berries, flowers all look somewhat similar. But a closer look reveals the difference in the arrangement of the leaves and the colour of the flowers and the berries. Every part of the Persian lilac is supposed to be poisonous, although I have seen bulbuls, especially the red-whiskered ones, enjoying the berries on several occasions.

I must mention the magic of the unfolding poppy! Today morning when my husband and I carried our tea outside, I noticed how pretty the scarlet poppies were looking, especially a plump bud that grew in the planting bed right next to us. But when I casually glanced at it five minutes later, it had opened halfway through, and by the time we finished our tea, it had blossomed fully. I'm sure it is a fairly common phenomenon, but it was the first time I witnessed something so awesome, my previous experience of seeing a flower bloom being restricted to one of those slo-mo videos.

15th

Overcast since last evening; a layer of beautiful cirrus fibratus clouds, stretched into delicate strands by the wind, is floating overhead today. This is one of my favourite cloud types, ethereal and otherworldly. Formed high up in the troposphere, these are ice crystal creations fashioned by strong continuous winds, but 'they tell nothing of the weather in store. Perhaps they are just there to look nice.'[4]

Nest building activity is in full swing. The yellow-flowered allamanda creeper outside our bedroom window was stripped of all its leaves by the cold, and even though new life is gradually starting to emerge, the bare boughs are a good hunting ground for nesting material. About 10 days ago, I noticed a female sunbird painstakingly peeling off thin fibres from one of the branches. A closer look through the binoculars revealed that it was a very skilfully executed job; the hard skin of the branch had been stripped off, revealing the soft inner part, which the bird removed layer by layer, filling her beak with as much material as it could hold. All the time serenaded by the male! A few days later, a pair of Oriental white-eyes had the same idea, but in this case, I noticed that both partners put in equal effort, toiling long hours to get the nursery completed.

18th

A very rewarding day. In the early afternoon, the kids and I, equipped with a camera and binoculars, went for a walk, our mission being the sighting of a pair of scaly-breasted munias that my son had photographed last week.

The adventure began soon after we left home when on a young neem tree in a quiet lane, we spotted a pair of Marshall's ioras and soon afterwards a pair of purple sunbirds. A tailorbird, in full breeding regalia, was singing his heart out, turning this way and that, showing off his smartly cocked tail with its elongated feathers, desperately trying to woo an uninterested looking female perched on a branch nearby. Many years ago, in Nimach, Madhya Pradesh, I observed a pair of tailorbirds building their nest in a tecoma creeper growing along the wall of our front porch. The male had been an all-year tenant of the creeper, sleeping behind a large leaf every evening, and so when the time came for him to start a new family, he faithfully chose the same landlord. Both he and his partner toiled industriously to stitch two big leaves together and then to line them with soft material. Wanting to help them, I left some fluffy cotton wool on the railing of the porch near their nest one morning and was delighted to find it gone by evening. I can't be sure whether they used it to cushion their nest or whether the wind bore it away, although I would like to believe the first scenario.

> The Tailor-bird's nest is a work of art. The thread, after being passed through the edges of the leaf and drawing them tightly together, is knotted at both ends so as to keep the sewing in position.[5]

Anyway, to come back to the present. The munias were where my son had seen them last week, foraging in a narrow grassy divide between two rows of houses. Right in the middle of the divide is a huge imli tree, underneath which is a permanently leaking tap. The birds treat it as their private watering hole, and on this sunny afternoon, there were silverbills, red-vented and red-whiskered bulbuls, a pair of Indian robins and their cousins, the magpie-robins, plus the ubiquitous mynas. On the higher branches of the imli was a flock of yellow-footed green pigeons, warming themselves in the late afternoon sun. A pair of silverbills was feeding on the ground, digging and probing with their beaks, while a second pair frolicked in some adjoining low-height bushes. I wonder if these are the birds that frequent our bird feeders. A male robin, his tail cocked jauntily, strutted about on the road before crossing over to the house right next to where we were standing. There he dived under a bougainvillea creeper and emerged triumphantly a moment later, holding a bright magenta flower in his beak. He then flew to a letter box a couple of houses away, where his lady love was waiting with some nesting material in her beak. For a fanciful moment, I imagined that caught up in the magic of spring, he would give the bloom to her. But no, he slipped through the narrow opening of the letter box most adroitly, and deposited it there, helping to prepare a colourful nursery for juniors. The moment he came out his partner went in to further cosy up the nest, and so the process of life carried on.

But by far, the most entertaining event of the day was the case of the jilted bulbul! In a young neem tree some houses away from the robin's nest, a red-vented bulbul was hard at work trying to woo a female perched on a branch just above his. He puffed up his feathers and metamorphosed into a fluff ball, all the time performing a funny dance that involved beating his wings up and down rapidly. The female seemed rather uninterested at first, but just as I felt that she was finally glancing his way, there was trouble in paradise. A rival male, flying furiously, alighted on an overhanging branch and proceeded to scold our friend most agitatedly. When he got no response, and the dance continued unabated, he bent down and gave the dancer a vicious peck with his beak. Caught unawares, the victim half rolled, half fell onto a lower stem, before gathering his dignity and flying off without so much as a backward glance, which was just as well, because the newcomer and the lady were now cosying on the same branch, wing to wing, beak to beak, thus leading us to the conclusion that the puffball was in fact the intruder!

Hugely entertained, we continued with our walk, with the children managing to photograph a lesser whitethroat and an ashy prinia. The lesser whitethroat is a winter migrant to Delhi, arriving in October from its summer grounds in Europe and departing back in March. It is a rather difficult bird to spot, confining itself largely to the canopies of trees, where it flits from branch to branch restlessly in typical warbler fashion.

Our final destination was a stretch of uninhabited government land, which had been fenced and allowed to grow wild. On a small raised platform just inside the main gate, someone had scattered some grain, and on this were feasting peacocks, Eurasian collared doves, pied mynas, bulbuls and parakeets. The peacocks with their glorious tails, resplendent in hues of turquoise and midnight, strutted about trying to impress the nearly half a dozen peahens. On a small rocky outcrop some distance away, we spotted a grey francolin, but being extremely shy, the bird had vanished into the undergrowth even before the camera could be adjusted. However, we were rewarded on the way back by a close-up view of a yellow-footed green pigeon in her nest. The silly bird had made the nest considerably low, down in a large bougainvillea creeper, and to me it seemed as though she had miscalculated the size as well, barely fitting into it herself!

20th

Spring equinox, and my favourite weather; capricious and temperamental; overcast with the promise of rain one minute, and the sun peeping out from behind the clouds the next. Sitting outside at the club for lunch with my boy; it was very still, with marbled grey clouds and the special liquid light that accompanies such days, which are dark and bright at the same time. Beautiful, indeterminate, sitting-on-the-fence sort of weather, unable to decide whether it wants to be sunny and spring-like or overcast and cold or both.

> *'It was one of those March days when the sun shines hot and the wind blows cold: when it is summer in the light, and winter in the shade.'*
>
> —CHARLES DICKENS, *Great Expectations*

We've had glorious weather this entire winter, with extended periods of cold, sunny days, interspersed with rain. Winter rain is magical and nothing lifts my spirits more than waking up to a dark, rainy morning, hearing the thunder roll in the distance, whilst cuddled up in my quilt. Winter rains in Delhi are caused by western disturbances, which are basically mid-latitude cyclones or low-pressure systems originating in the Mediterranean Sea.

Spring equinox officially marks the start of spring in the northern hemisphere, although I feel that the short Delhi spring sets in by the middle of February when the days become balmy and the garden becomes a riot of colour. At the two equinoxes, the spring and the autumn, day and night are of equal duration (well nearly) all over the world. What this means is that the sun is directly overhead at the equator at noon on these days. Fun fact: this is also the time of year when the sun takes the least time to sink below the western horizon at sunset and to rise above the horizon at sunrise.

Spring equinox is celebrated all across the northern hemisphere as the herald of new beginnings, a celebration of rebirth and renewal. It is also the start of the Hindu calendar, the Shalivahana Shaka and the Bikram Sambat, both of which emphasize the lunisolar cycle. The year starts on or after the spring equinox, in the month of Chaitra or March.

21st

Suddenly there is a lot of activity in the garden; birds flitting around, bugs getting about their work, butterflies hovering over flowers. The garden itself is a blowsy overgrown riot of blooms, probably its swansong before the growing heat makes everything wilt. Lots of cabbage-white and red Pierrot butterflies fluttering about in the garden these days.

Love is in the air and birds are pairing up and looking for nesting sites. A pair of red-whiskered bulbuls attempted to make a nest on the wall fan in the tiny veranda outside my room, and I had happily reconciled to doing without cool air for the next couple of months. However, the silly birds had not taken into account the slope of the fan and within a week, the nest was down. The silverbills, who are at the feeders the entire day, have started taking a break from eating to woo each other. I see the male inching closer, hop by hop, to the female sitting on an overhanging branch several times a day. This little dance often culminates in their flying off together into the sunset. The white-eyes are very active too; unusually so in fact. At least two pairs are attempting to find suitable nesting sites in the garden, and the pom-pom or calliandra bush growing all along the front wall seems to be their favourite hunting ground. Good thing I refused to get it pruned. If there is one grouse I have with Indian gardeners, especially the sarkari ones, it is their all-consuming desire to cut things down. And their understanding of pruning is to mercilessly chop-chop. As an example, notice the jarul trees at the roundabout. As soon as the flowering season is over, they will reduce them to stumps, ignoring the age-old wisdom of cutting back to just above a growing point, be it a bud or a branch. The trees will flower in the next season, but they are a mere shadow of their true potential. I find comfort in the fact that I'm not the only one to be bothered by the overzealous behaviour of the gardeners. In 1905, EHA, a very popular naturalist of his times, wrote in his wonderful book, *A Naturalist on the Prowl*[6]:

> Now, too, we plant roadside trees. We plant them, and when they have grown up and begin to be of some use, we hack them down again, because the rain dripping from their leaves is supposed to damage the roads.
>
> I once remonstrated with a *muccadum* whom I saw superintending this work. I told him that even native Governments had always regarded the planting of roadside trees, to give the weary traveller shade, as an act of piety and an antidote to the sins of previous births, and that he was wickedly undoing a good work which had cost much public money. He replied, "Not so. I do not destroy the roots of the trees. I only cut away the branches which spoil my roads." I said, "O! I understand. It is the roots that give shade to the traveller."

23rd

Coming back to spring and its temptations. The bird feed put out in a corner of the garden, which normally attracts entire flocks of sparrows, bulbuls and silverbills, now sees fewer visitors, partly due to the availability of alternate food sources (figs such as goolar and peepal are fruiting) and partly because the birds are too busy pairing up and exploring nesting sites. Our resident purple sunbird, whom I have seen evolve from a dull yellow eclipse plumage into a brilliant purplish-black, has found a mate and together they are inspecting potential shrubs to build their nest in.

> The Sunbird's nest is a cleverly built structure… It is a little oblong or pear-shaped pouch made of fine fibres hung from the tip of a branch in a bush or small tree. A great deal of cobwebs are used in strengthening the fabric of the nest… Most of the building work and the brooding is done by the female, but the male always keeps in the vicinity and accompanies her back and forth, constantly cheering her up by snatches of a merry little song.[7]

I had planted lots of freesias and gladiolus, and monitored their progress regularly, but alas! I was not vigilant enough, because the monkeys pulled out all the bulbs and ate them and now only one pot of each remains.

For me, by far, the sweetest herald of spring is the magpie-robin. Every morning, even before it is light, he starts his sweet trilling just outside my window and brings to mind the famous saying:

> *'Faith is the bird that feels the light*
> *and sings when the dawn is still dark.'*
>
> —RABINDRANATH TAGORE, *Fireflies*

25th

The wind is languorous today, lazily pulling the cotton candy clouds into punctuation marks, my favourite being an exclamation mark formed by a straight line cloud complete with a fuzzy dot at the end! I am reminded of one of my favourite lines:

> It was an ideal spring day, a light blue sky, flecked with little fleecy white clouds drifting across from west to east. The sun was shining very brightly, and yet there was an exhilarating nip in the air, which set an edge to a man's energy.[8]

Lots of bird activity this morning and in the half-hour in which we had our tea, the fruiting peepal was visited by three types of bulbuls, the coppersmith and brown-headed barbets, yellow-footed green pigeons, treepies and crows, pied and common mynas, parakeets, white-eyes, hornbills and, last but not the least, rosy starlings, who are visiting us on their way back to their summer grounds. A short while ago, a magpie-robin was singing from his favourite perch in the goolar, and purple sunbirds were frolicking on the balcony.

About a fortnight ago, on a visit to Shahpur Jat, I saw a female sunbird hard at work building her pendulous nest. I thought the choice of location, on a sickly-looking shrub right in the middle of the parking lot, ill-conceived, and unfortunately I was proved right. When I went back with my son two days later, there was no sign of the bird. The nest was there, hanging precariously, but it was obviously abandoned, with absolutely no activity. All in all, love's labour's lost, and we had to drown our disappointment in large glasses of cold coffee!

28th

Falguni Purnima! This is the full moon of Shukla Paksha in the Hindu calendar month of Phalguna and is associated with the festival of Holi. It is also the last full moon of the Indian lunar calendar. The festival of Maha Shivratri is also celebrated in the month of Phalguna. For the Buddhists as well, Falguni Purnima is especially holy because this is the day when Gautam Buddha was reunited with his family after many years; after Buddha received enlightenment, his father sent a messenger to ask him to visit the family, which he then did. In the West, this full moon is called the worm moon, probably due to the emergence of earthworms in the spring.

In the late evening, we were returning from a walk, and as we turned a corner, there it was, an enormous, orange-hued full moon, dominating the horizon! It might have seemed like the largest moon I had ever seen, but actually it is a phenomenon called 'moon illusion', whereby when the moon is low on the horizon, our brain compares it to nearby objects like trees and decides that it is massive. In contrast, when the moon is overhead, even if it is at its brightest, it is dwarfed by the endless vault of the heavens!

29th

A beautiful day! Pewter grey sky, the golden shower of neem leaves, a brisk wind and eagles gliding overhead. After lunch the clouds cleared, driven to other lands by the wind, and the afternoon was so bright that I was tempted to squint against the piercing light. The sky is a startling shade of Prussian blue, seldom seen in hazy Delhi. Sunlight glinting on the newly minted leaves of the peepal turns them a deep shade of rose gold, and try as I might I am not able to capture this extravaganza of colour on my phone, even though it has an excellent camera. Some part of the magic goes missing the instant I try to grab this fleeting moment, as if reminding me that some pictures are best left hanging in the hallways of the mind, where memory polishes them to an even higher lustre.

31st

Endings and new beginnings. The winter garden is literally going to seed, while all those plants that are sensitive to the cold and have been lying dormant and bare are bursting into new leaves. The nag champa and swarna champa have become green and leafy overnight, as has my frangipani. The mogra bushes, growing in a neat line on one side of the garden, are flowering, and my husband likes to collect the blooms every morning. The star jasmine creeper and the raat ki rani or night-blooming jasmine bush now fill the dusk with their heady fragrance. Lots of butterflies in the garden; large cabbage white, common Mormon, common wanderer, common jay, red Pierrot. On a visit to the outskirts of Delhi, the wheat is golden in the fields, ripe for harvesting.

April

*'The sun blazing fiercely,
the moon longed for eagerly,
deep waters inviting
to plunge in continually,
days drawing to a close in quiet beauty,
the tide of desire running low:
scorching Summer is now here, my love.'*

—KALIDASA, *Rtusamharam, Canto I: Summer*

1st

April is the start of the really hot season in North India. Gone are the balmy March evenings which define the liminal edge between spring and early summer. Gone too is the garden umbrella, relegated to its corner in the spare room once again. An average high of 39°C, which often crosses the 40° mark and a low of nearly 25°C is hardly conducive to outdoor pleasures. The cloud cover is steadily decreasing through the month, while the length of the day is increasing, leading to longer clearer days and more heat. The only bright spot is that the relative humidity is not very high, settling in at an average of 20 per cent or so. However, nature, unlike me, seems to love the onset of summer. Most trees are clothed in shiny new leaves, the gulmohur and amaltas are starting to bloom and the birds are getting ready to rear their young. Other flowering trees this month include the beautiful jacaranda and the palash. A flowering tree that superficially resembles the palash is the Indian coral tree, which is also at its superlative best this month, bearing two-inch long flowers that grow in dense terminal clusters. The flowers are a flamboyant shade of scarlet, appearing all the more striking because they emerge in March and April when the tree is leafless. It is also called the tiger's claw tree because of the black spines/thorns that grow along the branches. The siris as well as the Krishna siris are in bloom as well, the canopy dotted with golden puff balls, making for a very pretty sight. The siris is an interesting tree, both when in flower and later when the long, flat, flaxen-coloured seed pods form in November.

5th

Hearing a loud commotion in the garden, I peeped out through the window to see a treepie, who had come to drink water, being mobbed by a pair of red-vented bulbuls. They obviously had a nest nearby and didn't trust that the purpose of the treepie's visit was to only quench his thirst.

The lantanas are flowering and attract a host of butterflies; common jay, wanderer, yellow orange tip, pioneer, cabbage white and for the last few days, common lime and common Mormon as well. They animate the garden, bringing it to life, as they flit from shrub to shrub. With their amazing butterfly-attracting abilities, lantanas deserve more respect and popularity than they currently enjoy. They also provide a very welcome burst of colour at a time when the winter garden is winding down and the beds are looking bare and forlorn. The two allamandas are in bloom once again, their plump buds almost as pretty as the blossoms themselves. The jasmine creeper that I planted just outside our bedroom window a few months ago has grown some more, and once the nasturtiums that smothered it the entire winter had died down, it almost immediately burst into flower, its incredibly sweet fragrance drifting into the room at dusk.

6th

Every season brings with it endings and fresh beginnings, and at no time of the year is this more apparent than in spring. Late March and early April sprinkle magic fairy dust everywhere, and suddenly there are signs of renewal and rebirth. Foliage is the main showstopper of April, rivalling more flamboyant flowering trees and shrubs. The three mussaendas that have lain dormant all winter, reduced to mere sticks, are in new leaf, and I can see the colourful flower bracts forming; red and white and pink. The red passion flower creeper has spread across the fence. I hope it flowers soon! Today I spotted the first blooms on the blue pea or aparajita creeper, although the vine itself is still rather sorry-looking. I have planted some zinnias, portulaca, cosmos, sadabahar and gaillardia.

8th

Delhi is aglow these days! There is no other way to describe the beauty of the different fig trees; pilkhan, anjeeri, peepal. Fourteen varieties of figs grow in Delhi, and each one is prettier than the other, especially during this season. But the crown goes to pilkhan, a large, spreading, deciduous tree, which develops a gorgeous golden canopy in the spring, with leaves ranging in colour from red to purple to russet. I find the latter most beautiful of all. All the colours of autumn in spring, but with a difference. The browns and russets of autumn, although beautiful in their own way, are of old leaves, which no longer have the translucence of new growth. The golden leaves of the pilkhan are young and shiny, and when light falls on them, they glow magically. A shiny, glowing canopy of bronze and gold!

A very striking fig in new leaf this month is the banyan or badh, with most trees also fruiting simultaneously. Starting life as an epiphyte, because of which it is also called strangler fig, mature banyans send down aerial roots, which help them to spread laterally, often over a wide area. One of the biggest banyans in the world is located near the town of Kadiri in Andhra Pradesh. Estimated to be nearly 550 years old, it covers almost five acres and can fit 20,000 people in its shade![1] Among the other venerable banyans located across the country is the iconic banyan in the Botanical Gardens at Kolkata, estimated to be nearly 250 years old, and a large tree called Kalpabata in the Jagannath Temple in Puri that is estimated to be even older.[2,3] The banyan is sacred to both Hindus and Buddhists, and there are many references to it in the literature of both religions. On a more mundane level, the pretty red figs are a favourite with birds, and a fruiting tree near our house has a steady stream of visitors all day.

> We owe an account of how it came to be called the 'banyan' to Thomas Herbert, who travelled through Persia in 1627-29. One such tree grew close to the present-day port city of Bandar Abbas, where 'banians' (banias or traders) from western India had decorated it with ribbons and built a temple in its shade. Therefore, says Herbert, it has been 'named by us the Bannyan Tree'.[4]

The laurel fig is another amazing tree, with a huge canopy and a comparatively much shorter trunk. But what it lacks in length, it makes up in width, the latter feature getting accentuated by the aerial roots which wrap around the top of the trunk. The new leaves of spring stand out beautifully against the older, less glossy ones.

The kosam or lac tree is arresting in spring, visible even from a distance. The new leaves, which emerge in March, are a beautiful, glossy scarlet red, a shade not commonly seen in the trees of Delhi. The leaves will remain this colour till the end of April when the red will gradually turn into a bright shade of green. The flowering coincides with the emergence of new leaves in March–April, but the small bright yellow blooms are no match for the brilliant red leaves and are totally overshadowed by the latter.

9th

A beautiful cool morning filled with birdsong, although the day will most likely become hot and oppressive. The peepal is clothed in new leaf, in dazzling shades of pink, peach and green, and seems to have developed a wide, spreading canopy overnight. There is much bird activity; a pair of treepies is noisy and confiding, coming very near to where I'm sitting for a drink of water from the terracotta pot hanging in a corner. I am delighted and feel privileged. A pair of mynas are trying to decide on a suitable nesting site from amongst the two available on a side branch of the peepal. Eventually they choose one and take turns tapping and widening the hole till one of them flies off and returns with a shiny piece of paper that it promptly deposits inside. Over the next hour, they bring in feathers, plastic and a long scrap of printed fabric. The plastic, being light, keeps flying off and is recaptured after dexterously executed mid-air sallies.

Our three resident species of bulbuls—the red-vented, red-whiskered and the least common, the white-eared—flit about the whole day. They are joyous, energetic little birds, weaving through branches in pursuit of their prey, turning and twisting at impossible angles. They are bold and confiding and regulars

at the bird feeders. I find that they like to perch on the very top of bushes and small trees, and the tecoma creeper is one of their favourite haunts, as is the bougainvillea I planted some years ago. Unfortunately, their social mannerisms leave much to be desired, and this week alone I have seen the red-vented ones tumble to the ground fighting like mynas on three occasions. The red-whiskered and the white-eared bulbuls, however, seem to be less pugnacious, with the former being my absolute favourites. I find them to be very pretty, with their crests and the red in their cheeks. They are also exceptionally melodious, and have a lovely warbling song plus a range of other calls. But what I like best about them is their playful nature, darting from shrub to shrub, chasing after each other, dancing about the whole day.

> Bulbuls do more to keep the world lively than any other bird I know of. They do not sing outside the pages of *Lalla Rookh*, but they have sweet voices and light hearts, and they seem to bubble over with a happiness which is infectious.[5]

A flock of babblers visit the potted plants every afternoon, and I confess to being slightly confused about the purpose of their stopover. They hop from pot to pot, poking around for insects, often settling in the mud to cool themselves, and by the time they leave, it is as though a mini storm has swept through the place.

The magpie-robin is at his showy best in the summer, singing his heart out from his favourite spot on top of the lamp post by the side of the bamboo fence.

10th

For the past month or so, our neck of the woods has been taken over by rosy starlings. They are everywhere, in almost every fruiting tree in the neighbourhood, their pretty pink and black bodies glistening in the sun. Their lively 'tchi-tchi' can be heard all through the day, first from the peepal and the chamrod, which are full of berries, and then as the day progresses and becomes hotter, from the shady neem. They are restless birds and the entire flock will suddenly fly out of the tree and wheel around for a minute or two before settling down again, in an incredible synchronization of swoop and sweep. I wonder sometimes whether they rise up following the up-baton of a hidden conductor and waltz around for a charmed moment, before alighting again at the same spot!

Rosy starlings are passage migrants through Delhi, making a stopover both to and from their wintering quarters in Gujrat and peninsular India. They breed in East Europe and Central Asia, and are among the first migratory birds to arrive—reaching here by August–September—and the last to leave in April. They certainly don't seem to have any special attachment to their summer grounds, spending most of the year in their wintering grounds or on the wing.

11th

A very cute thing this morning. The peepal adjoining our house is laden with berries and is host to a variety of hungry birds all day. A rufous treepie flew onto one of the mid-rung branches and almost immediately gave a short but insistent call in his customary guttural tone. At once his mate, who was feeding on the nearby goolar, flew down, and perching herself beside him, accepted gracefully the berry he put into her mouth before flying off again. This is true love; plucking a berry off a laden branch and feeding the other, who is incidentally, perfectly capable of helping herself to the fruit!

I have always thought of the treepie to be somewhat of a quarrelsome bully, an opinion reinforced when many years ago I saw one plunder a dove's nest. But apparently, no one is immune to the charms of the nesting season, which in the case of the treepie lasts from March till July, when the voice may continue to be harsh but the heart is soft!

Food sharing between mated pairs of birds is called allofeeding, and is often a part of the mating rituals.[6] It is thought to be a way of nutritional enhancement and of course, pair bonding. Allofeeding also refers to unrelated adult birds feeding fledglings; I have often observed several adults in the flock of jungle babblers that frequent our garden feeding titbits to the same juveniles.

12th

The two goolars near our house are fruiting heavily, and the ground underneath is covered in red semi-squished berries, which have a very distinct odour. A pair of crows have their nest on the smaller of the two trees, and one parent sits guard on a nearby branch all day, largely whiling away the time because the resident mynas and bulbuls pose hardly any threat. The only time the bird is spurred into some action is when there is a koel nearby, or when the black kite, who is on a nest-building mission herself, flies past. At once, the crow gets into fight mode, and setting up a raucous cawing gives the intruder chase, flying wing to wing, settling down only when the larger bird is at a safe distance.

The kite, on its part, is unconcerned; it has more important things on its mind, like choosing the perfect twig for its nest. To accomplish this, it first does a recce, circling overhead a couple of times to decide on the exact stick it wants. After a moment spent deliberating, it does a fly past and snaps the branch off effortlessly, the entire manoeuvre so efficient and well timed that I have yet to see a miss, even when the bird has had to swoop really low between the trees to get to something lying on the ground.

The nest, high up on the neem, is a sturdy-looking, albeit untidy bowl made from twigs and other assortments. Scrutiny with binoculars revealed two empty packets of India Kings cigarettes and a shiny inner wrapper from one of them woven through the mesh!

Crows and koels do not see eye to eye; the former may be smart but the latter are smarter. Koels are brood parasitic and lay their eggs in the nest of the house crow; the breeding season of the two overlaps, lasting through spring to the monsoons. The eggs also bear a resemblance to each other, although those of the koel are slightly smaller, thus hatching a little earlier. There is much cleverness, and underhandedness involved in the manoeuvre, with the female koel being known to remove the eggs laid by the crow and replace them with her own, while her mate distracts the host.

13th

Baisakhi! Baisakhi falls on the first day of Vaishakh, the second month of the Hindu calendar, and is observed as the start of the traditional Solar New Year in many parts of India. Falling on the 13th or 14th of April, this time of year marks the end of the agricultural cycle for many crops, and Baisakhi is, therefore, also celebrated as a harvest festival. Baisakhi has special significance for the Sikhs, this being the day when Guru Gobind Singh started the Khalsa order.[7]

A tree which catches my attention every day while on a morning walk in Nehru Park is the Caribbean trumpet tree. For most of the year it is entirely ordinary and unremarkable and not worthy of a second look. But come March, and it breaks into the most glorious, flashy golden yellow flowers. Seeing them silhouetted against the clear azure sky of a cloudless morning makes me rethink the combination of bright blue and bright yellow, which I have never really been fond of. It is a gorgeous, uplifting sight, especially since the tree is totally leafless, leaving only a golden mist to dazzle the spirit.

16th

The putranjiva or wild olive tree is covered in pale, lime-green new leaves. When backlit by the sun, it is a magical sight, and any new growth that might get ignored normally stands out as a tiny pinpoint of light. This is an evergreen tree native to more moist forests and is slightly stunted in Delhi because of our comparatively drier climate.

I love the peepal at this time of the year; the new leaves are showy and tender and glisten as though polished to a high veneer. Changing in colour from pink to copper before turning green, they give the term 'wearing new clothes' a whole new meaning, and as sunlight sifts through the tree just outside our gate, dappling patterns on the cobbles below, I notice the design change almost daily as new leaves emerge and the canopy fills up. The pretty heart-shaped leaves also make the most amazing music, sensitive even to the slightest breeze. The berries ripen in mid-April and usually grow in pairs. Ranging in colour from green to dark purple when ripe, they appear to be glued directly to the stems; I certainly have never been able to spot a single stalk on any!

> In some ways peepal trees are great show-offs. Even when there is no breeze, their beautiful leaves spin like tops, determined to attract your attention and invite you into their shade. And not only do they send down currents of cool air, but their long slender tips are also constantly striking together to make a sound like the pattering of raindrops.[8]

Another gorgeous tree to see in March and April is the moulmein rosewood, very aptly called 'jewels on a string'. While walking in Nehru Park, I always like to linger under one that grows near a bend in the track. The tree is leafless in March for a brief period, before the most spectacular, mauve flowers take over all available space. The flowers appear to be ombré hued, ranging from light mauve to a darker purple, and droop in long grape-like sprays.

20th

By the middle of April, the garden is stripped down to its bones, so to speak. The exuberance of winter blooms is over, and any stragglers have been pulled out. The beds look suddenly bare, brown earth cracking under the midday heat. I am tempted to plant some more summer annuals but decide instead to let the land rest for some time. I am not happy with the quality of the soil; it looks thin and undernourished and pockets of it are overrun with termites. So my first task is to feed it huge amounts of vermicompost and neem khali (I don't like to use chemical pesticides). Bought some more periwinkle or sadabahar, but I have had limited success with growing these in the last couple of years, so fingers crossed they do well this time. Healthy flowering plants suddenly shrivel up and die, leaving me perplexed and distressed. However, thankfully, the gaillardia and gomphrena are doing well, as are the portulacas.

A flock of quarrelsome pied starlings—or pied mynas as they were called when we were young—on the gulmohur are squabbling loudly, but making a very pretty picture against the beautiful flowers, which have completely enveloped the canopy. Eventually, the victorious pair, after driving away the rivals, settles on a branch in the very centre of the tree, making plans and acting amorous, while the losers shift to the garden, pecking at titbits, chasing insects, sipping water, bathing and generally creating a ruckus. A very vocal pair of purple sunbirds are flitting about, feeding on the last of the late blooming salvias; the male in full wedding attire is an arresting sight, his blue-black metallic plumage glinting in the sun. A tangle of old leafless branches at the very top of the allamanda continues to be a favourite hunting ground for nesting material for the smaller birds, white-eyes, sparrows and purple sunbirds, who methodically peel off the soft inner fibres layer by layer.

22nd

A light dust storm in the afternoon and the evening sky is a moody shade of rustish-red, probably because the thunderclouds are still high enough to reflect the red-orange light of sunset. Red light has a longer wavelength than blue, and so at sunset, when sunlight has to traverse a longer path to reach us, the red is still visible while the blue has been mostly absorbed or removed. Since clouds contain droplets that are bigger than light waves, they are able to scatter light without changing its colour, and voila, take on the hue of the incoming light themselves!

Coming back from Khan Market in the late evening after celebrating my daughter's birthday, I notice that the sky is overrun by fruit bats, their black silhouettes picture perfect against the orange sunset. Lutyens' Delhi is home to a large number of these flying foxes, as these big bats are also known. Most of them roost in the area in their favoured old avenue trees of arjun, jamun, mango and ficus, especially those growing near a water body. Here they spend the whole day lazing, oblivious to the noise and chaos of the busy roads, flying out at dusk in a seemingly never-ending stream. Their diet consists of ripe fruits and nectar, making them great pollinators and seed dispersers.

However, we don't need to travel far to see them; our two goolar trees are their favourite late-night snack addas!

> Of all the wild-fowl included under the name of bats, the only one that really comes into the foreground of Indian life is the fruit-bat or flying-fox. This animal has

what I consider a handsome face, with large soft eyes, and would not be a bat at all but for two characteristic points, a strong batty smell and an insatiable craving for strife. Flying-foxes carry this last trait further than any others of the tribe.[9]

26th

A gorgeous supermoon. Rising low on the horizon at dusk it has a distinct reddish hue and appears bigger and brighter than usual. A full moon becomes a supermoon when it coincides with the moon's perigee or the point in its orbit when it is the closest to the Earth, making it appear nearly 14 per cent larger than usual.

This is the pink moon of the northern hemisphere, so named because of the pink-coloured phlox that blooms in North America at this time of the year. In the Hindu calendar, this is the Hanuman Jayanti full moon of the Lunar month of Chaitra and is considered especially auspicious. It is also the first full moon of the Hindu New Year. For the Buddhists, especially in Sri Lanka, this is Bak Poya, celebrated as the day when Buddha visited Sri Lanka and brokered peace among warring factions.

Since this has been an unusually pleasant April, it has been possible to linger outside till dusk and I have been able to observe the moon every day, to puzzle over its rising fairly early in the evening, even while the sun is still up. This mystery was eventually solved by a very interesting book by Tristan Gooley, where I learnt that a young moon rises earlier than sunset, getting delayed each day till the full moon rises very nearly at sunset, lingering below the horizon after that.[10]

27th

Our son champa (*Michelia champaca*), which I planted a few months ago and is barely four feet high, is flowering; the orange-yellow blooms are highly fragrant and very pretty. It did cross my mind that if I were to pluck one and bring it indoors, it would perfume the entire room, but I couldn't bring myself to do that to my tree.

Son champa has several uses; its timber is used to make high-quality furniture and the flowers are used for worship in temples and also as a base for perfumes. In fact, the son champa is also called the 'Joy' tree because Jean Patou's famous perfume 'Joy' uses the essential oils distilled from its flowers as a base![11]

28th

March onwards, the chamrod tree outside my son's room is full of small white flowers, which attract tiny insects, which in turn attract birds like chiffchaffs, bulbuls and white-eyes. Later, towards the end of the month, the flowers turn into pretty orange berries, which are a favourite with rosy starlings, rose-ringed parakeets, barbets and bulbuls. However, till about mid-April, the tree is hijacked by the rosy starlings, who use it as a sort of day roost, their noisy chattering obliterating any other sound. It is only after they have left for their summer quarters that the other birds become noticeable and audible! A rose-ringed parakeet comes to feed every morning and follows the same routine of pottering around for a few seconds before bending down to pluck a small bunch of berries, which he then transfers from his beak to his claws. He enjoys his breakfast, finishing off the berries one by one, spitting out the ones he doesn't like, before helping himself to another bunch.

Yesterday evening saw much activity in the goolar, and I fetched the binoculars to investigate. A female rose-ringed parakeet was sitting on a branch, eating berries from a large bunch held in her dainty feet, while three other females hovered around like bees drawn to a pot of honey. Every few seconds, one of them would inch closer diffidently, and from a safe distance, stretch out her neck as far as she possibly could, hoping to have a go at the berries. This audacity was dealt with swiftly by the rightful owner, and for a couple of minutes, order would be restored. But soon, one of the other birds, getting impatient and unable to wait anymore, would launch an aerial strike, hoping to impound the treat by force. Much shrieking and shouting followed, ending only when the first bird, having had her fill, dropped the remaining fruit and flew away, but not before one of the others had grabbed it. I didn't stop to watch any further but am sure the same drama would have unfolded once again. Why the birds couldn't just fly the minuscule distance to the chamrod tree to get their own berries, I don't know! Birdbrained creatures!

29th

The garden has a storybook feel to it today. A short rain shower in the morning has made the day surprisingly cool. Periwinkle and zinnia bloom, birds at the feeders, emerald green grass strewn with leaves, doves cooing on the windowsill and butterflies flitting about dreamily. A flock of more than a dozen pioneer butterflies dance about the garden the entire day, attracted by the lantanas that are flowering profusely. April showers are very welcome in a rapidly heating urban landscape. Every colour seems brighter and more saturated against the grey of rain clouds; in the evening, noticed the crimson petals of the gulmohur against a leaden sky and thought about the many hues of April; Prussian blue sky, shiny lime green of new leaves, the silver oak clothed in gold and silver, the beautiful lilac of the flowering African mahogany, bright scarlet of the kosam.

My silverbills have babies and I'm delighted the parents brought them to the garden. Today early afternoon, I saw parents and two young ones on a branch of the allamanda right next to the feeders, sitting all in a row, sticking to each other. The fledglings are very noisy and keep up a constant chatter whilst they are being fed.

May

The days, hard enough to bear, get longer and longer. On the first of the month, the sun rises at 5.40 a.m. and sets at 6.56 p.m., giving us more than thirteen-and-a-quarter hours of hell. By the end of the month, the sun rises 16 minutes earlier (5.24 a.m.) and sets 17 minutes later (7.13 p.m.), adding more than half an hour of unwelcome daylight.[1]

1st

May is the start of the uncomfortable period of 'heat and dust' in North India. The days are scorching, with an average high of 42°C; not a leaf moves, not a cloud offers respite. The sun beats down relentlessly, endlessly, day after day after day, making this the month with the maximum sunshine (an average of 12 hours). In 2009, Delhi recorded its highest-ever temperature in May at a terrible 52°C. The days are almost windless, and the hot, still afternoons can get very oppressive. To make matters worse, thunderstorms become frequent as the month progresses, covering everything in a patina of dust. I have always conjectured that May is considered to be one of the windiest months in Delhi only because of these thunderstorms, the days otherwise being absolutely breezeless. It rains for only one or two days in the entire month on average, and with the cloud cover remaining at a scanty 8 per cent, stepping out in the afternoon is like putting your head into a sweltering furnace. However, despite the lamentable weather, May paints the city with broad brush strokes of saturated colour; the yellow of amaltas, deep red of gulmohur, pink of jarul and lilac of jacaranda dazzle against a blue sky!

May is also the start of the mango season in Delhi. The few mango trees in our locality are covered in fruit, unripe and green as yet, and the target of the rose-ringed parakeets, who will most likely leave very little to be harvested.

2nd

May belongs to the amaltas tree. A flowering amaltas has to be one of the most glorious sights in nature; cluster upon cluster of drooping yellow flowers, varying in intensity from light to very dark, from cream to lemon, to the more common bright yellow blooms smothering every branch. My favourites are the beautiful ombré coloured bunches, with the lighter flowers usually at the top. I love the way the petals drift down slowly with the slightest breeze, blanketing the roads in golden snow, and find it amazing how this nondescript tree, which has a largely untidy appearance during the rest of the year, comes alive this month and shouts, 'Look at me!' Entire avenues are filled with this gorgeous golden flower, as are roadside parks. An interesting folklore—in Kerala it is believed that the monsoon arrives 45 days after the tree bursts into dense flowering!

> Few Indian trees are more beautiful when in flower. Draped in streaming clusters of bright yellow blossoms, which hang from its branches in a golden shower, the tree suggests the European Laburnum, but it is infinitely more beautiful.[2]

Called the 'aragvadha' or 'disease killer' in Sanskrit, the amaltas is mentioned in various ancient texts. The pulp of the fruit has laxative properties while the root is said to be useful in managing heart problems. My grandfather, who lived in Chandigarh, insisted that the cylindrical seed pods were a panacea for all his digestive problems and always picked up a couple of ripe black ones on his morning walk. Even the leaves possess medicinal properties and a poultice prepared from them is used for joint pains and rheumatism and for soothing insect bites. The beautiful flowers have their uses too, being antipyretic and anti-inflammatory.

All along the Shanti Path, amaltas alternates with the equally striking gulmohur; a beautiful tree at this time of the year, bright red against a bright blue sky, the flowers showy and five-petalled; four of them flame coloured, ranging from orange-red to scarlet, while the fifth, which is the largest, is white speckled with red and yellow and brings back memories of many a childhood hour spent searching for the perfect specimen.

The mahua tree is clothed in new, dainty pink leaves, which will gradually turn dark green as they age. This is a very important tree of tropical mixed forests of central, eastern and peninsular India, where almost every part is used either commercially or as a food item.

4th

High drama today morning whilst we were having our tea outside in the garden. Hearing a commotion I looked up just in time to see a crow fly past at full speed, followed closely by a pair of black kites. In and out they weaved through the branches of the peepal and the saptaparni. My husband noticed that the crow was holding what looked like an egg in its beak, and immediately the pieces fell into place! The kites have a nest high up in the neem across the road, and for the whole of last month they have been hard at work, collecting nesting material, which is no easy task for a bird of that size. A flypast over the area is the first requisite, followed by identification of the desired twig, then a brief strategizing regarding the best angle to reach it while on the wing, and finally a swoop, engineered so swiftly and smoothly, effortlessly almost, that I am nearly convinced the target has been missed. But such is the bird's dexterity that it always emerges triumphant, holding its prize aloft, and on a couple of occasions I have seen its partner relieve it of its burden mid-flight, the twig exchanging beaks or claws efficiently.

But coming back to the thrilling chase. What probably happened was that the crow, always an opportunist, seeing the nest unattended for a moment, trespassed into forbidden territory and stole the egg. The parents were furious, and rightly so, but the crow, taking advantage of its smaller size, managed to give them the slip by hiding in the dense foliage of the goolar! This is probably the second clutch of the season for the kites, because last month the children and I observed, on two consecutive days, a smartly liveried juvenile sitting on the lamppost right outside my room, its spots and stripes gleaming with the brilliance of youth.

Last week I saw a crow attempting to cross the road opposite the milk booth with the most enormous twig in its beak, which was much too heavy for the bird to fly with, hence the road journey. However, after a few minutes, it very sensibly gave up the struggle and flew off to a nearby tree, where it probably had its nest.

6th

The courting treepies are building a nest in the goolar tree, their chosen branch a considerable way up, and at an angle from the main canopy. It is well camouflaged in the greenery, and I discovered it only because of the constant comings and goings of the birds. Not that they try to keep their location a secret; they are noisy and do not hesitate to scold any trespassers. Just this morning, I saw them give a crow a proper chase and then harshly tick off a pair of mynas who had ventured too close. In the initial stages of the nest-building activity, both birds were equally active, darting to and fro with suitable material. But as the nest begins to take shape, I notice the female doing the actual construction work, staying on site, arranging the twigs with dexterity, while her partner flies about procuring the raw material. Every once in a while, she takes a break and they sit together on a nearby branch, romancing, rubbing beak to beak, the male even feeding her on occasion. In the afternoon when it is too hot to work, I see them come to the water pot hanging in a shady corner of the garden, usually one at a time, but sometimes together as a pair. I am glad I bought a larger-sized terracotta pot this summer, and now both birds can perch on its sides comfortably.

8th

For the past several days, I have been noticing that many avenue and roadside trees have 'gone to seed' and are overhung with pods of all shapes and sizes, the comparison of which makes for an interesting study and helps while away an hour or two of the endlessly long afternoons.

The most conspicuous of these is the subabool, a common roadside tree. The flowers, in a very light shade of cream, are tiny and grouped together into small balls. They resemble the flowers of the mimosa, and when unripe look like small green peas glued together! The brown-coloured pods are long and flattened and hang together in clusters, easily distinguishable even from a distance. While some trees still have a few flowers, the majority are covered with seed pods this month. There was a time, more than two decades ago, when the subabool was considered a miracle tree because of its use as firewood and as livestock fodder. However, on the negative side, it is also one of the most invasive trees and even becomes a pest in some areas.

The shisham is another roadside tree, which shrugs off its cloak of anonymity come March when the new pale green leaves make for an arresting sight. The flowers are short-lived and unremarkable, but the small flat pods that follow soon after are interesting. Starting out green, before turning a pretty shade of beigish-brown, they hang together in bunches, their grouping making them noticeable.

The pods of the siris are flaxen in colour and smother the

tree in a tawny haze. They are long and flat and keep up a constant chatter all day. In contrast, the pods of the kassod are a darker shade of brown, and are not as flat as those of the siris; the undulations in their surface are caused by the seeds and are clearly visible. Their colouration, ranging from pink-red to green and brown, confused me initially, but I learnt from Pradip Krishen's book that the side of the pods facing the sun takes on a deep red-pink hue, while the opposite side remains green.[3]

The amaltas, which has painted the city golden this month, will soon develop its dark-coloured cylindrical seed pods which are reported to have numerous health benefits. The pods of the maharukh or ulloo tree, which, from a distance, looks very similar to the neem, are thin and papery and hang in yellowish-green clusters when young. The pods of one of my favourite trees, the moulmein rosewood, also form in May, but are unremarkable, being an indistinct shade of brown and shaped somewhat like an un-filled balloon, only bigger. On the other hand, the pods of the earpod wattle are quirky; curly and wrinkly and greenish-brown in colour.

The jacaranda is a beautiful tree and tints entire avenues purple. The pods, which will appear next month, are small, flat and circular and of a lovely brown colour. They keep the tree looking interesting long after the flowering season is over. The arjun tree also has an unusually shaped seed pod; oblong and faceted; it ripens to a pretty dark brown colour. The kapok, which towers over most other trees, has a capsule-like pod, which when ripe—usually by the end of May—splits open to release clouds of white cotton wool-like fluff into which the seeds are embedded. The semal or silk cotton tree also fruits at the same time, and like the kapok, its pod ruptures to cast its downy fibre, with the seed attached at one end, into the air. As children, we would chase after the magically drifting wisps of white, hoping to catch one! Though superficially similar in appearance, any confusion between the two trees ends in March, when the semal burst into the most gorgeous, showy red flowers. The kapok also flowers in March, but its blooms—although very pretty in themselves—are not as striking as those of the semal, being smaller and green in colour.

The seed pods of the babool, which also ripen in May, give the impression of a string of pearls hanging from the branches. An innocuous tree otherwise, of an average size, with an average canopy and mean-looking thorns, it comes into its own during the rainy season, when it is completely smothered in small yellow pom-poms! The fruit pods of the jhand, a tree native to Delhi, are long and cylindrical and twisty, curly, much like mountain roads with their unexpected turns! They ripen from May onwards, changing in hue from green to brown. The flowers of this tree are also interesting; tiny pale yellow blooms, which cluster together in long spikes, much like a bottle brush but without the handle!

Most interesting of all, however, are the aptly named pods of the jungle jalebi or the monkey pod tree. The curly-wurly pods, which form from April onwards, actually look like a jalebi, especially when they ripen to a deep red. They remain interesting even after they split open, with black seeds hanging from red threads. The pods are edible and believed to have numerous medicinal properties, as do the leaves. The kanju or Indian elm also has a very curious pod; a papery round disc, very thin, with a seed embedded in the centre, which starts out green before ripening to a beige brown by May.

10th

The silver oak is nearing the end of its flowering season. It is such a beautiful sight; silver and gold and green, although the trees in Delhi seem to be a stunted version of the ones that are found in the Himalayan foothills, growing alongside jacaranda and banjh oak in strands of mixed forest.

The silver oak is a tall, evergreen tree, with a canopy that grows straight, rather than spreading out. The undersides of the fern-like leaves are shiny silver and contrast beautifully with the bottlebrush-shaped golden orange blooms, which are a favourite with birds like bulbuls, mynas and barbets. This is one of the fastest-growing trees in the world and is very popular in tea and coffee plantations, where it is used to provide just the right amount of shade to the young plants.

12th

A beautiful tree that flowers this month, the jacaranda, has the most gorgeous, fragrant blue-purple blooms, arranged in drooping clusters. Blue flowered trees are not very common in Delhi, so the explosion of violet is a treat for the eyes.

Called neel mohar or neeli gulmohur locally because of the resemblance of its leaves to those of the gulmohur, this native of South America and Central America gets its English name from the South American Guarani language, in which the word means fragrant. Pretoria in South Africa is known as 'Jacaranda City' because of the more than 70,000 trees lining its avenues and dotting the parks and gardens in spring.[4]

14th

For the past several weeks we have been serenaded morning and evening by our resident koel. He comes to the same branch of the same saptaparni tree at exactly the same time every dusk and predawn. In the morning, half asleep, it seems to me as if he's serenading the dawn, while at the same time bidding farewell to the night, singing in the liminal zone between the two. Add the music of the magpie-robin to the solo of the koel, and you have the most melodious aria of all. I can think of no better way to start and end the day!

An unusually large number of butterflies these past few weeks. Hordes of tiny grass blue; both tigers, striped and plain; cabbage white and Indian cabbage white; pioneers and common wanderers; common jay and common Mormon.

> 'These Butterflies, in twos and threes,
> That flit about in wind and sun –
> See how they add their flowers to flowers,
> And blossom where a plant has none!'
>
> —W.H. DAVIES, *Flying Blossoms*

18th

A very pleasant May so far, overcast with the promise of rain; or windy and sunny; or my favourite of all, an actual downpour. All of nature is rejoicing at this unexpected gift. May is the hottest and driest month of the year in Delhi, and any break from the scorching heat is welcome. The grass has turned a gorgeous shade of emerald green; the gardenia bush is blooming profusely under my bedroom window, with the fragrant white blooms twinkling at dusk; the tecoma bush has bright yellow flowers; the sita-ashok is clothed in new pink leaves; and in between the showers, the sky is a pretty shade of eggshell blue, streaked across with high drifts of white clouds. Nature, ever resilient, has put up a show of technicolour brilliance!

I planted a sita-ashok tree a few years back, which after struggling for several months initially and nearly withering away, has responded to my tender loving care and grown to a respectable size. Although it is still young, it has started to bear a few flowers, and the new leaves, which appeared this month, are adequate recompense and make the waiting for proper blooms worthwhile. The young leaflets, borne on drooping branches, are a gorgeous shade of copperish-pink and remain tenderly folded for quite some days before opening up and gradually turning green.

A flowering sita-ashok is a wondrous site; clusters of small orange, scarlet and vermilion flowers crowd every branch and stem, and are a perfect foil for the dark evergreen canopy. Lightly fragrant, each flower has numerous stamens, which give the cluster a slightly ethereal, blowsy look. The sita-ashok has a long association with ancient Hindu and Buddhist literature and culture. When Sita was abducted by Ravana and taken to Lanka, she refused to live in his palace, preferring instead to stay in a grove of ashok trees (Ashok Vatika), thus giving them her name forever. Even today, in present-day Sri Lanka, the Hakgala Botanical Garden is situated in the area known as Seetha Eliya or Sita's Garden, which has the Sita Pokuna, a barren hilltop where Sita was held captive. The Sita Amman temple is located nearby, as is a stream where Sita is supposed to have bathed.[5] Buddha is said to have been born under a sita-ashok tree at Lumbini. Sanskrit texts give at least 16 names to this tree, and Kalidasa, in one of his plays, writes that the flowers appear only after a young maiden kicks the tree in spring.[6] The Hindu God of love, Kamadev, was said to keep a flower of the sita-ashok (one amongst five different blooms) in his quiver.

20th

The nest of the treepies is complete, the eggs have been laid and the mamma bird is in place. The eggs were laid about 10 days ago, in a nest that initially appeared to be too flimsy and small, but was able to withstand a strong thunderstorm a few days ago. The wind howled and the rain came down in sheets and I was sure that either the branch or the nest or both would be on the ground before long. Surprisingly, both were able to withstand the onslaught remarkably well and the birds have since been diligently doing their duty. Except today, when the male was found to be rather wanting and negligent.

Around 7 o'clock in the morning, while we were having tea in the garden outside, I did my usual recce with binoculars and found all peaceful in the treepie household. The female was sitting on the eggs in her customary fashion, slouching so low as to be nearly invisible, with only her tail with its pointed edge giving away her presence. Five minutes later, the male flew down to the nest, they exchanged greetings beak to beak and the female flew off, probably to eat her breakfast and take some well-deserved rest. For some time after that, her partner stayed dutifully perched on a nearby branch, but before long, getting bored, he flew to an adjoining tree. I was surprised; leaving the nest unattended seemed a risky thing to do, what with our resident black kite always in the vicinity. And sure enough there was trouble almost immediately. An opportunistic crow, sensing his chance, made a dive for the nest, landing on the very same branch, and it was only the immediate response by the male that saved the day or rather the eggs! He dive-bombed the enemy furiously, shrieking all the time at the top of his voice. The intruder flew off, but the treepie now suitably chastened, decided it would be best if he sat on the eggs himself, which he did in a very clumsy fashion; half sitting, half standing awkwardly, not crouching efficiently like the female. Fortunately for him, she came back shortly afterwards and giving a cry of delight, he stepped out of the nest immediately, giving her back her rightful place. Observing the treepies for more than a month has led me to realize that they have '…a large variety of calls, some loud, harsh and guttural, others pleasing and melodious; one of the commonest of the latter being a clear *bob-o-link* or *kokila*.'[7]

21st

Gang wars! We have our resident flock of grey babblers, who put in a noisy appearance every morning, exactly at the time we have our tea, with the specific intention of sharing our Marie or Digestive or whichever biscuit we are having that day. Today, however, there was trouble in paradise in the form of a rival flock of babblers, who insisted on partaking of the treats, despite warning shouts from the home team. A free-for-all ensued, with bird fighting bird and tumbling to the ground very near where we were sitting, momentarily oblivious to our presence. It was amusing to see the battle, with nearly all the adults participating and the juniors striking up a cacophony from the fence.

24th

The koels start up their insistent duet at four every morning, maybe even earlier, before the light starts creeping across the sky announcing daybreak. Our bed is right next to the window and in the hush of the night, their song washes over me in an outpouring of sweetness, and I lie in bed,

waiting to hear which bird will join in next. It is usually the magpie-robin, followed by the bulbuls and sunbirds, so that by the time we have our tea, the chorus outside has reached its crescendo; a symphony of melody and out-of-tune songs, with the babblers joining in and also the black-rumped flameback woodpecker, who pays us a visit almost every morning, coasting in with its distinctive undulating flight and settling on one of the saptaparnis, poking around the bark for insects. A famous American poet has the following to say about woodpeckers:

> 'His Bill an Auger is
> His Head, a Cap and Frill
> He laboreth at every Tree
> A Worm, his utmost Goal.'
>
> —EMILY DICKINSON

The koel sings again at dusk, continuing long after it gets dark, its song my personal dawn and twilight. But not everyone shares these sentiments, certainly not 'ESS', who says of the koel[8]:

> Next is the Koel, whose 'ku-oo' grows more strident with each successive rise in scale until you feel you can no longer bear it. But just then his voice cracks, his clamour breaks off abruptly and you find yourself positively rejoicing in the thought that the wretch has choked himself. And just then he begins all over again.

25th

The maulsari tree is in bloom, clusters of creamy-white flowers illuminating the dense canopy like stars spread across an inky sky. They fill the evening with their heady perfume before falling to the ground with the advent of day. The fruit of the maulsari is small and oval-shaped, ripening to a bright red, making a very pretty picture interspersed with the flowers. The leaves, with their wavy edges, are easy to identify. The flowers of the maulsari are used in making attar.

The figs of the goolar are also ripening, turning from green to a bright orangish-red. They are a favourite with the fruit bats, who visit the tree at night, leaving behind a considerable mess to be cleared up in the morning.

26th

A Buddha Purnima, super, flower, blood moon lunar eclipse! Phew! That's quite a mouthful. At 8 o'clock, the moon had risen above the horizon, and was an otherworldly red in colour, bathing the sky around in a pinkish orange glow, like that of the sun at sunrise.

Sunlight bends and scatters as it travels through the Earth's atmosphere, but the light at the blue end of the spectrum scatters more than at the red end. Usually we see red light scattered only at sunrise and sunset, giving the sky its rosy glow, but during a lunar eclipse, some of this light reaches the surface of the moon, and I love the way NASA's page has correlated this scattered red light with the red glow of the moon. 'The eclipsed moon is dimly illuminated by red-orange light left over from all of the sunsets and sunrises occurring around the world at that time.'[9]

Being at perigee, the May full moon this time is also a supermoon, appearing bigger and brighter than usual. The Native Americans called it the flower moon because of the abundance of spring flowers at this time of the year.[10]

Buddha Purnima or Vesak is celebrated on the full moon day of the month of Vaisakha in both the Buddhist and the Hindu calendars. It holds special significance for Buddhists, this being the day when the Buddha was born and also attained enlightenment.

28th

The moringa or sonjna tree in our neighbour's garden has been fruiting for the last two months, providing us with a steady supply of drumsticks and consequently a ready excuse to make sambhar every other day. The bean-shaped drumsticks, also called sonjna ki phali, are immature seed pods, plucked while still soft and light green in colour. Mature drumsticks have their uses too; the seeds can be removed from the pods and roasted to make a delicious nutty snack.

Every part of the moringa tree has health benefits, giving it the moniker of 'miracle tree'. The leaves are the most nutritious, followed by the pods. Dried leaf powder is commonly sold as a dietary supplement. The leaves, pods and seeds also have antidiabetic and anti-inflammatory properties.[11]

> Both the leaves and fruit of the sonjna are an astonishingly rich source of calcium, iron, vitamins B, A and C (when raw) and of protein. One scientist calculated that ounce for ounce, sonjna contains the calcium contained in 4 glasses of milk, the vitamin C equivalent of 7 oranges, the potassium of 3 bananas and more than 3 times the iron found in spinach.[12]

Another miraculous quality of this tree is that it can grow in less than ideal conditions; in arid regions and areas of poor soil. Also the leftover seed cakes can be used to purify drinking water!

June

'Dry-throated, foaming at the mouth,
maddened by the sun's sizzling rays,
tuskers in an agony of growing thirst,
seeking water, do not fear even the lion.
Peacocks, exhausted by the flame-rays of the sun
blazing like numerous sacrificial fires,
lack the will to strike at the hooded snake
thrusting its head under their circle of plumes.'

—KALIDASA, *Rtusamharam, Canto I: Summer*

1st

The terrible heat continues, and just when you think it can't get any worse, the temperature goes up by a couple of degrees centigrade—43, 44, 45, then 46. There is little that can redeem these scorching days. The average number of rainy days is 7, and the length of a day is nearly 14 hours. It is true that the jarul is in flower, the kochia and vinca in my garden are blooming, fruit shops are laden with mangoes, the birds are raising their families and nature is going about its business as usual. But for me, this is a time for hunkering down inside, curtains drawn and the AC on; like the ipomoea in the garden which wilts every afternoon, so do I!

2nd

The last day of Nautapa, or the nine hottest days of the year, during which it is believed that if it neither rains nor do cool winds blow, the monsoon will be a good one. The nine days commence when the sun enters the Rohini Nakshatra. Usually these nine days are scorching, but this year it has rained twice during this period, which, although offering a very welcome respite from the heat, may not augur well for the monsoon.

> Shortly before the monsoon, the heat becomes very intense. It is said that the more intense it becomes the more abundantly it will draw down the rains, so one wants it to be as hot as can be. And by that time one has accepted it – not got used to but accepted; and moreover, too worn-out to fight against it, one submits and endures.[1]

4th

A very welcome wet start to the month. Went to the nursery in the evening; humid but pleasant, which is a bonus for this time of year, when the days are usually scorching.

Nothing new at the nursery; bought a pink ixora and a magenta flowering plant that the maali at the nursery insisted was a new variety of hibiscus! But the highlight of the evening were the birds; flocks of scaly-breasted munias feeding on the wild grass growing in a large unkempt plot of land at the back of the nursery, a pair of hornbills excavating a hole in a dried tree stump, a very noisy party of alexandrine parakeets, and finally, peacocks. At least 10 of them, of which the one nearest to where I was standing, barely a few feet away, put up an incredible dance performance!

Birding is so serendipitous an activity; the most magical sightings happen when you are least expecting them. Like a few years back, when we were travelling from Kausani to Chaukori (in Uttarakhand) we stopped to have chai at a small shack on a lonely stretch of road, and there right in front of us, on the moss-covered branches of an adjoining banjh oak, were two families of woodpeckers, parents and juveniles—one scaly-bellied and the other Himalayan.

> Birds will give you a window, if you allow them. They will show you secrets from another world—fresh vision that, though it is avian, can accompany you home and alter your life. They will do this for you even if you don't know their names—though such knowing is a thoughtful gesture. They will do this for you if you watch them.[2]

5th

On the way back home from the market, I noticed that the jamun trees were flowering. The deep purple, slightly astringent, berry-shaped fruit will ripen by month's end and become an instant magnet for birds and young children! The jamun is closely interwoven into the fabric of Indian mythology. The fruit is said to have formed a very important part of the diet of Lord Rama, Lakshmana and Sita while they were on their 14-year exile. The Puranas narrate that Jambudweep, the island of jamun trees, which was located at the heart of the world, was the only abode of humans. Krishna's dark skin is compared to the shiny luminescent jamun, which in Hindu mythology is the incarnation on Earth of the God of the clouds, Megha. Hence the storm-cloud coloured fruit! The leaves of the jamun are still used in pooja and to decorate the entrances to temples and houses. Ibn Battuta, the 14th-century explorer, chronicles in his *Travels in Asia and Africa 1325–1354* that Delhi was full of jamun

trees, the fruit of which 'resembled an olive but sweet'.[3]

Every part of the jamun tree is used in Ayurvedic medicine. The small purple-hued fruit is a powerhouse of nutrients and is used as a cure against diabetes, liver and heart problems and various other illnesses. I remember my grandmother collecting the seeds of the jamun after we had eaten the fruit—washing, drying and then grinding them to a powder to be had daily as a preventive against diabetes.

The evergreen jamun is a common avenue tree in Lutyens' Delhi, where it lines entire roads, forming shady pathways. Ancient trees, with their spreading canopy, are the favourite roosts of the fruit bats.

6th

Rain and thunder in the afternoon for the second day in a row. I love the rain and I'm delighted but worried about the effect this may have on the timing of the monsoon. A crow pheasant poking about in the old bamboo fence, looking for insects. The pom-pom bush or calliandra has started flowering and attracts our friends, the sunbird couple. A flock of hariyals or yellow-footed green pigeons on the saptaparni. Red-whiskered bulbuls come looking for the crumbs of bread I have scattered for them. On a drive earlier today, I noticed a flock of cattle egrets in breeding plumage, sitting on the topmost branches of two adjoining trees, turning the canopy a bright shade of orange!

There is a lull in the rain and the sun comes out, making me imagine that there is a rainbow somewhere; although hemmed in by an urban landscape, it is difficult to spot one. But then I imagine that it is always raining somewhere, the skies are a cloudless azure somewhere else, and rainbows forming a beautiful double arc straddle across valleys far away. There is magic at work in nature!

> This grand show is eternal. It is always sunrise somewhere; the dew is never all dried at once; a shower is forever falling; vapor is ever rising. Eternal sunrise, eternal sunset, eternal dawn and gloaming, on sea and continents and islands, each in its turn, as the round earth rolls.[4]

8th

Our resident myna pair has become distressingly aggressive, chasing away all birds that dare to venture into their territory. Their modus operandi is clear; they will advance threateningly towards the intruder, hoping that their stance will convey their disapproval. This strategy usually works with smaller birds, like sparrows and silverbills and even bulbuls. However, if the trespasser is made of sterner stuff, and continues to feed nonchalantly, he is liable to be given a vicious peck in the back. This is accompanied by harsh scolding, much fluttering of wings and short aerial dives. The pair drove away the babblers and even a crow this morning. Even the squirrels don't faze them. I like the garden to be full of birds and am wondering how to put an end to this deplorable behaviour.

> Here you have a gentleman who makes the best of both his worlds—the wild and domestic. A carefree jungle roamer, he nevertheless perches on a ledge of the bungalow and nests in the roof. And woe betide any who encroach on the domain to which he has staked claim. Observe him

when he shelters from the noonday sun on a niche in the porch. Assuming a comfortable, digestive attitude he will commune in scarce audible tones with the lady of his choice. Lacking an appreciative feminine audience, he will talk to himself.[5]

10th

Today our resident juvenile shikra (he appears to be in the post-juvenile transitional stage now), who visits the water pot every day, often staying perched on the edge for half an hour or more, actually plonked himself right into the water for a good 10 minutes; not bathing or fluttering his wings, just sitting quietly, probably enjoying his cool dip. The heat is getting to all of us! On an overhanging branch a treepie kept up a constant scolding, which I found somewhat surprising. However, it was the babblers, the saat bhai, the alarm clocks of the avian world who, as usual, heckled and drove away the poor shikra, chasing after him till he was out of sight.

I notice that the shikra, after some time, moves onto standing on one leg. I find this puzzling because according to conventional wisdom, birds tuck one foot in their belly to reduce the loss of heat on a cold day. But this is the hottest part of the year and surely the shikra has no desire to hold onto body heat. Maybe it is more comfortable to shift feet when standing for a long time or maybe this is a sort of semi-relaxed pose, where the bird is alert but resting.

11th

Very hot; heat smothers the city in a stifling haze, the eerie stillness broken only by mid-afternoon, when the loo begins to blow. Loo is the hot and dry summer wind that blows across North India during the months of May and June. This dust-laden, scorching wind sweeps in from the arid northwest, from the Thar Desert, which forms a natural boundary with Pakistan, where the contiguous desert is called Cholistan. Loo can lead to heatstroke and even prove fatal, and it is advisable to stay indoors during the hot afternoons when it is at its peak, and maybe drink the cooling sherbets of khus and rose, although I personally prefer lassi. Before the use of air-conditioners became almost mandatory, homes used to have coolers fitted with screens of the fragrant grass, khus or vetiver, which when kept damp actually proved to be a very effective chilling agent.

13th

A violent dust storm in the evening, kaali aandhi, brought some relief from the terrible heat, but the respite it offered was in no way commiserate with the mess created. Trees uprooted, branches broken, dust swirling in eddies, an ominous red sky, absolutely low visibility with the dust blocking out the sun, and after the storm had passed, hours of cleaning to do at home. And worst of all, the rain that accompanied the squall, lasted no more than two minutes.

17th

Incandescent sky, wind like a lick of fire, but nature is resilient, and despite the terrible heat, the garden is flourishing. The sadabahars or periwinkles are doing exceptionally well—touchwood. The flowers are big and in showy shades of pink and orange-coral, dark gulabi, wine, white with pink centres, magenta with dark plum and the sweetest of all, the small fuchsia-coloured ones, which are a favourite with the red Pierrot and tiny grass blue butterflies. As ever, the ones planted in the ground in the bed right next to our tea spot are doing much better than the ones in pots. The gomphrena and portulaca and moss roses make a very pretty picture against the emerald green grass, the latter two a huge magnet for the bees, who hover around the entire day. The flowers, in cheerful, vibrant colours, open with the first rays of the sun and bring the garden alive. The zinnias are doing better than the previous years, but somehow I always struggle with them. The yellow cosmos have also not done well this time, and make for a very sorry picture, the stems weak and sickly looking and the flowers small and ill-formed.

A lovely surprise this morning when we went out with our tea. A male koel, dark and glossy, and a large green barbet at the water pot together. The koel had his drink first and jumped onto an overhanging branch, after which the barbet quenched his thirst. This is the first time that I have seen a barbet come to the garden for water this summer, and although he was very fidgety and on edge, he did take a few sips. The intense hot and dry summer is tough on all creatures.

19th

If there was ever a system of allocating to each month its most evocative symbol, June for me would belong to the magpie-robin, singing his wonderfully melodious song perched on the bamboo post supporting the gardenia bush outside my window. The song, which starts with the first light of dawn, is so mellifluous and delivered with so much passion that it is able to hold its own against the higher-pitched notes of the koel, and later the cacophonous calls of the mynas and the babblers. As the month has progressed, my little bird has got himself a mate and, to my delight, the said bamboo has become their favourite rendezvous place.

> Sweetly does the Magpie Robin sing in the small hours of the morning, when we are in our beds, but if you want to know what he can do, look at him and listen to him as he follows "the fair, disdainful dame" and his rival from branch to branch and tree to tree, suffering the ecstatic pains of a jealous suitor. What a masher he is in his new spring costume, with his black and white tail expanded like a fan, and his glossy breast at the very point of bursting with the frenzies of song which spout and gush from his swollen throat![6]

20th

Terrible, unrelenting heat, which coupled with the humidity, has converted the city into a sauna. The only movement is of the clouds as they come and go, raising hope before dashing it to the ground again. The official date for the arrival of the monsoon in Delhi is 29 June and I do hope it doesn't get delayed.

Over the last few days, I have been noticing that many of the palm trees in our vicinity are either flowering or fruiting. The royal or bottle palm—so named because of its resemblance to a two-toned bottle, the lower two-thirds of which is a smooth grey, and the remaining one-third a bright green—is in flower, its interesting blooms clustered together in tightly packed branches, which always make me think of bunches of wheat sheaves tied together with a string. The flowering stems are enclosed together in tubular sheaths, which burst on maturity to release the sprays of blossoms. As a child, I remember being very intrigued by this tree, after a gardener told me that if I inscribed my name on the trunk, it would endure forever, rising higher as the tree grew in height.

The wild or sugar date palm, on the other hand, is fruiting; the semi-ripe berries hanging in big clusters are noticeable even from a distance, essentially because of their bright orange colour. This is the palm of the delicious gur (jaggery), one of my favourite winter treats. Palm jaggery or khajoor gur, obtained after boiling the sap tapped from just under the leaves, is made mostly in the state of West Bengal, where it is used both as a dessert and a dietary supplement.

But by far, one of the most interesting palm trees of Delhi is the jaggery or toddy palm, the leaves of which appear to have been chomped off by an animal, so that they have the look of raggedy-edged triangles. Each part of this incredible

tree is useful; the leaves are used for making baskets; the trunk itself contains a starch similar to sago; the fruit hanging in huge tassels is edible; the sap is used to make a sugary treacle and to prepare toddy and arrack or palm wine.

21st

Summer solstice! The sun is directly overhead the Tropic of Cancer, making it the longest day of the year in the northern hemisphere and the shortest in the southern hemisphere. On this day the sun travels the longest path through the sky and reaches its highest point. However, it is not the day with the earliest sunrise of the year or the latest sunset. The former occurs a few days before 21 June, and the latter a few days later.

The summer solstice has traditionally been celebrated as Midsummer's Day by many countries in the West, with some even taking it to be the official start of the summer season. But I rejoice for a different reason. For us in North India, it marks the end of the scorching summer and the start of the monsoons. I look forward to the gradual shortening of the days—although that takes a long time to happen—and the onset of rains and cooler weather.

In Hinduism, 21 June marks the start of Dakshinayan, or the southward movement of the sun, which lasts for six months and ends with Makar Sankranti or Uttarayan in January. In Hindu mythology, these six months are the night-time of the gods.

It has been overcast the whole day, with the sun attempting to cast a gander every once in a while. But the evening has seen the most amazing mackerel skies; rows upon rows of undulating altocumulus clouds, stretching right up to the horizon. I remember the old rhyme 'mackerel sky, not twenty-four hours dry'[7] and wish it proves to be true for us in Delhi. These clouds were thought to resemble the scales of a fish, hence the name. Another interesting lore has to do with high-altitude cirrus clouds, colloquially called 'mare's tails' because of their resemblance to the said horse tails. If the sky was overrun with both kinds of clouds; patchy altocumulus (mackerel sky) and wispy streamer-shaped ones (mare's tails), it meant that bad stormy weather was on its way.

'Mare's tails and mackerel scales make tall ships carry low sails.'[8]

This divine meteorological warning gave sailors, out on the open waters, time to lower their sails and head to the safety of the shore.

23rd

One of my favourite trees, the kadamba, is flowering, the adorable orange-yellow flower balls striking against the dark foliage. The stigma of the tiny individual flowers protrudes from the tight clusters, giving the golf-sized balls a fuzzy look. Sacred to both Hindus and Buddhists, the kadamba finds mention in the *Bhagavata Purana* (between AD 800 and 1000) and in Jayadeva's *Gita Govindam* (AD 1200). In North India, the tree is associated with Krishna and Vrindavan, and there are numerous accounts of Krishna playing his flute or performing his miracles in the shade of this fragrant tree, giving it the name of haripriya. Kadamba is also considered holy in South India, where it is supposed to be the favourite of Goddess Parvati and many festivals and legends are woven around it. In Madurai, the tree holds special significance as 'sthalavruksham' or tree of the place, and a withered relic is preserved at the Meenakshi Temple.[9]

The kadamba has many uses; the flowers are used in perfumes and attars, while the leaves are said to be effective in controlling diabetes. The wood is easy to work with and is used in the paper and pulp industry, and also to construct boxes and some furniture components.

24th

Today morning saw the happy closure of the episode of the baby myna, which started yesterday evening at around six when there was a sudden commotion in the peepal outside our house. A baby myna had fallen from its nest in the tree and was standing completely still, frozen with shock, while its parents, fearing the stray dogs who live in the parking nearby, had set up a loud, alarmed chanting. We wondered what to do; leaving the fledgling there would have most certainly meant it becoming dinner for either the dogs or the pair of kites who live in the vicinity. Eventually, deciding that it was our only option, we brought the fledgling home in an old shoe box and kept it on a table on the first-floor terrace. Almost immediately, its parents flew in and settled on the terrace wall. It was dusk by now and we shut the door and pulled the curtains in a bid to encourage feeding and family consultation. The children had crumbled up a biscuit and put it in the shoe box, but the baby didn't need it; its parents started bringing it food almost immediately. After an hour or so, fortified and stronger, the fledgling hopped out of the shoe box onto the table and in the morning when we opened the door, we saw it fly off with its parents. It was obviously not hurt, just shocked by the fall, and a night of rest had done it good.

26th

For the past couple of days, clouds have started rolling in by late afternoon, and a sudden stillness descends upon the darkling day, sending a thrill of anticipation coursing through my body. But alas! The promise remains unfulfilled; a strong wind starts up almost immediately, developing into a full-fledged storm, an aandhi; the clouds are driven away and swirling dust obscures everything in sight. And most disappointing of all, the next morning dawns scorching and humid and stays that way till afternoon, when the entire dispiriting process starts up again. It seems that the ancient wisdom of the Nautapa is spot-on; any rain or cool winds during that period of nine days interferes with the monsoons.

An unfortunate fallout of this entire drama has been that I have not been able to observe the June full moon with any clarity, something that I was really looking forward to. The full moon closest to the summer solstice—also called the strawberry or rose moon—appears very low in the sky, because of which its light has to shine through extra layers of atmosphere to reach us, giving it a reddish-pink hue. Since this moon will be the closest to the Earth, it would have appeared much brighter than usual, and much larger too. And I missed it! For Hindus, the full moon of Jyeshtha (June) or the Jyeshtha Purnima has special significance for married women who celebrate it as Vat Purnima.

27th

Overcast, rainy, windy; a strange orange glow permeates everything—sky, air, the evening itself. My garden may not win any prizes, it being a little overgrown and wild, but it has become a haven for wildlife. Bees throng the moss roses and portulacas from morning to evening, butterflies hover over the gomphrenas and vincas, while bugs and ladybirds abound in the verges. The bird feeders always have a waiting list, with silverbills and doves and parrots squabbling over favourite spots, while bulbuls, babblers and sparrows prefer the grain scattered on the ground in a corner of the garden kept deliberately untended. My friend, the crow, comes the moment I sit outside with my tea, both morning and evening, demanding titbits from the biscuit tin, as do the quarrelsome mynas. It's a happy place, with white-eyes bathing and sunbirds fluttering over the allamanda and the pom-pom bushes, tailorbirds trilling and ashy prinias skulking. We are treated to melodious aria twice a day by the koel and magpie-robin, and with all this richness, who needs a prize?

July

*'Now come the days of changing beauty,
of summer's parting as the monsoon comes,
when the eastern gales come driving in,
perfumed with blossoming arjuna and sal trees,
tossing the clouds as smooth and dark as sapphires:
days that are sweet with the smell of rain-soaked earth.'*

—BHAVABHUTI, *The Rains*

2nd

July and August are the rainy months in Delhi. They are also the sensory months; a time for gathering clouds, and letting them drift through the imagination, for enjoying a hot cup of adrak elaichi chai with pakodas. These are the Hindu months of Shravan and Bhadrapada, the rainiest time of year. July and August also see the maximum cloud cover, with the sky being overcast more than half the time. 31 July is considered to be the cloudiest day of the year, and the chances of it being overcast are nearly 60 per cent. The monsoon hits the city in the last week of June, usually on the 29th, although the humidity starts building up days in advance, making July a very muggy month, with an average high of 38°C and a low of 30°C. The number of rainy days averages 16 while the humidity is high at nearly 60 per cent.

At the beginning of July, the chance of a rainy day is 43 per cent, which increases to nearly 55 per cent as the month progresses. During the month, the average length of each day decreases by 50 seconds, which means that on 1 July, the sunrise is at 5.26 a.m. and the sunset at 7.23 p.m., whereas on the 31st the sunrise is at 5.42 a.m. and sunset at 7.13 p.m., making for a difference of nearly 25 minutes.

3rd

Delhi has a mix of semi-arid and humid subtropical climates, although I personally think that it is the former that impacts the day-to-day weather more. The city lies in the scanty precipitation zone—areas having an average annual rainfall of 50–100 cm—with a very woeful average of 617 mm. That leaves me yearning for rain the entire year, and any build-up of clouds, however insignificant, makes my heart beat a little faster.

July in Delhi is the eternal wait for the rain, the quickening of heartbeats as the sky turns indigo, and then waking up one magical morning to leaden skies and a steady downpour. It is the droplets of water sparkling on the caladium after a night of rain, the cluster of dragonflies that hover almost motionless in the air, the stratus and stratocumulus clouds that paint the sky in shades of grey and pewter and charcoal, much like an artist who smudges her canvas with random strokes, and on windswept rainy evenings, it is the curtains billowing far into the room.

An interesting fact about July is the shifting of the wind direction exactly in the middle of the month, from the west (from the 1st to the 14th) to the east (15th to 31st).

5th

The sausage tree has been flowering for the last couple of months, but I feel that it is reaching its most prolific phase now. It is a handsome tree, with a large spreading canopy; deciduous in our dry climate, but evergreen in the moister areas of its native Africa. The most interesting feature of the tree is the manner in which the flowers and fruits hang down at the end of long, dangling, flexible, rope-like branches, which can reach a length of several feet. The large flowers, which are a distinctive shade of red-maroon streaked with yellow and green, are thought to give off a specific scent at night that is particularly attractive to fruit bats.

The shrubs of calliandra, which form a sort of hedge along our front boundary, are flowering, the blooms living up to their sobriquet of 'red powder puff'. The buds are as pretty as the puff-balls themselves, looking exactly like unripe raspberries. The main flowering period is from November to April, but I have been noticing a small second flush in the rains as well. The flowers are a favourite with purple sunbirds and parakeets, the former preferring to drop by when the sun is up, while the latter are early morning visitors. A few months ago, at the beginning of spring, a pair of white-eyes and red-whiskered bulbuls scouted around the dense canopy, searching for a nesting site, but nothing came of it, which is just as well because the leaves shut at sundown, leaving the branches suddenly bare and exposed.

The dawn chorus continues to bring joy to our mornings; the song of the koel followed by the melodious serenade of the magpie-robin; barbets joining in soon after, and then the lovely whistle of the red-whiskered bulbul. The tailorbird trills a very sweet tune and the doves coo amorously. A new addition to this orchestra is the impassioned call of the peacocks, which is more heavy-metal than a sweet symphony.

6th

I have hung two big terracotta water bowls for the birds, which witness a steady stream of visitors the entire day; crows, treepies, bulbuls, magpie-robins, mynas, sparrows, silverbills, parakeets, babblers, white-eyes all come to quench their thirst. The crows have started visiting in groups of four, waiting for their turn impatiently, hopping about on branches and in pots. I recognize them now, especially the pair that has a nest in the neem tree nearby, and whose male member has become my friend over time. He is the messier of the two, splashing water every time he dips his beak to take a drink. The female is daintier, both in appearance and in behaviour and only occasionally does a drop escape her beak. They usually come together or within a few minutes of each other. While they were building their nest, they would often drop the twigs they were carrying into the water before having a drink, whether to soften them or because they had no other option, I don't know.

The magpie-robin sings his heart out from his perch on top of the lamppost, and when his throat gets parched, comes by for a reviving sip of water. Last week, I saw a purple sunbird enjoying a drink, and for a moment was afraid that it would topple right in. But fortunately, no such catastrophe occurred and after frolicking around for a few minutes, he joined his partner on the allamanda creeper. I am always amazed at the volume of his song as compared to his size!

8th

July is the month of the frangipani or champa tree, which, although flowering for the entire summer, is at its most prolific at this time of the year. Also called the temple tree because of the tradition of offering its leaves in temples, this tree has a special place in Hinduism and Buddhism and is sacred to both religions.

> This is the tree so frequently cultivated in the neighbourhood of temples, where it supplies the continuous demand for flowers used as votive offerings to the gods. Its remarkable power of bursting into leaf and blooming even when taken out of the soil has led it to be regarded as an emblem of immortality.[1]

The flowers have a heady, sweet, lingering perfume, and often, especially on warm sultry evenings, their fragrance has reached me long before the tree has come into view. Delhi has two types of frangipani, although, within each type, there exists a startling diversity. The flowers of the *Plumeria obtusa* or white frangipani are, as the name suggests, white, but those of the plumeria rubra occur in a delightful pastel palette. White with yellow centres or yellow at the edges, pink and yellow ones, pink and white, or only pink or only white, red with yellow, or only red and even light orange. I have a beautiful mature frangipani growing in a big pot; it has light pink, almost blush-coloured flowers with orangish-red centres. During the rains, some of the petals turn a dark pink stippled with white and make for a very pretty sight. We also have a red champa; the gorgeous crimson flowers have small yellow throats and hang in clusters from a long red stalk.

10th

A blustery overcast morning, the kind when you can actually see the wind as it blows through the trees.

> *'Who has seen the wind?*
> *Neither I nor you:*
> *But when the leaves hang trembling,*
> *The wind is passing through.*

> *Who has seen the wind?*
> *Neither you nor I:*
> *But when the trees bow down their heads,*
> *The wind is passing by.'*

—CHRISTINA ROSSETTI, *Who Has Seen the Wind?*

Layers of cloud at noon, thin gossamer webs passing over the denser stratocumulus, much like lace curtains placed in front of heavier drapes. Suddenly the thunder rolls in and the rain starts; billowing curtains of white transforming trees and shrubs into a mysterious woodland dripping with enchantment, the darkling afternoon imbuing every leaf with a burnished patina. The birdsong has died down, and the only sound is the rumble and roar of the thunder as it passes through. I love rainy afternoons and wonder at the magic alchemy that makes louring grey clouds appear suddenly in an incandescent midsummer sky. One minute it is scorching, with the curtains drawn against the noonday glare, and the next moment you become aware of a change in the quality of light, a gradual darkening of the day. I am reminded of the beautiful poem, *Noboborsha* or New Showers, written in 1900.

> *'The clouds rumble, rumble high up in the heavens.*
> *The rain rushes in.*
> *The new stalks of rice quiver.*
> *Doves shiver silently in their nests,*
> *frogs croak in flooded fields,*
> *The clouds rumble, rumble in the heavens.'*

—RABINDRANATH TAGORE

12th

Weather lore has, since times immemorial, been an integral part of every culture and country, because in a world without the technological and scientific advancements that we take for granted today, reading the signs of the weather could have made the difference between life and death. Survival would have depended on being able to predict an incoming storm or the correct season to hunt or to grow crops, or to leave cattle out on the pastures to graze. Since India has been an agricultural economy from the very start, and the agriculture itself has been largely rainfed, most of the weather lore is centred around the timing and health of the monsoons. Each part of the country has its own weather knowledge, passed down the generations, with some of the observations having become proverbs over time.

In North India, the period in late May or early June when the sun enters the Rohini Nakshatra is called the Nautapa, nine days during which it is believed if the heat remains continuous and unrelenting, the monsoon will be a good one. I have tested this one myself and found it to be largely true, probably because the intense heat deepens the low pressure which pulls in the rain-bearing winds. There is an Odia folklore which states that rain starting on a Saturday continues for seven days, although my aunt who lived in a crowded gali of sabzi mandi in Delhi, and who was a virtual repository of ancient wisdom, insisted that it was only rain that started on a Wednesday which would continue for more than five days or more at a stretch. Another saying from Haryana and Punjab states that the hotter the month of June, the stronger the monsoon and I think this would have the same logic as the Nautapa. The direction the winds blow from during Holi and the festival of Akshay Tritiya also indicate the strength of the monsoons in North India. Winds from the north and west will ensure a good monsoon, while those from the east will indicate the opposite.

In other sayings, sparrows bathing in the dust, frogs croaking, peacocks calling insistently and ants carrying their eggs signify that the rainy season is here, as does a cluster of dragonflies hovering nearly motionless two metres or more above the ground (this one I have tested and found to be largely true). Lapwings laying their eggs during the night—although I am not sure how this could ever be verified—is set to portend heavy rainfall. Good foliage on the peepal and mahua trees during spring also indicates a good monsoon, according to weather lore prevalent in North and Central India. However, if crows start cawing at night and owls hoot during the day, it could bode a possible drought, as could rainfall on a sunny day.

Interestingly, even Kautilya, the master statesman and administrator, describes in his book *Arthashastra* (fourth century BC) the correct way to measure rain. He further tries to correlate the health of the monsoon to the visibility of Venus in the eastern sky during the monsoon season.[2]

15th

The two orange jasmine or andhra kamini shrubs growing in a corner, right next to our tea spot, are in bloom; the astonishingly beautiful and fragrant flowers adding sweetness to our morning brew. The white flowers are small and arranged in loose clusters, and will be followed by pretty red fruits, which look very much like rose hips. The star jasmine or juhi is also clothed in small white flowers this month, and I always think how aptly named it is. The white florets do look like twinkling stars, set against a dark green sky, but it is their

sweet fragrance that makes them a winner. A vine of the common jasmine (although I find nothing common about it) or *Jasminum officinale*, grows in a big pot on the balcony, twining itself dexterously around the jute rope that will help it climb all the way up to the roof. It has lovely five-petalled (again white) flowers that have a heady fragrance, which is at its best in the evening, reminding me of the following lines:

> *'From plants that wake when others sleep,*
> *From timid jasmine buds, that keep*
> *Their odour to themselves all day,*
> *But, when the sunlight dies away,*
> *Let the delicious secret out*
> *To every breeze that roams about;—'*
>
> —THOMAS MOORE, Lalla Rookh

The four mogra shrubs growing in a line next to the allamanda creeper have flowered profusely during the last two months, although now the frequency has reduced somewhat. Flowering occurring in phases is a common feature of the plant and often takes place less than a month apart throughout the summer.

Jasmine is grown for its highly fragrant flowers, which to my mind proclaim summer—with its perfumed sultry evenings—more than anything else. The flowers have traditionally been worn by women in their hair, especially in South India, although the trend is becoming increasingly popular in the north as well. The flowers are also distilled to make delightful jasmine tea. In some parts of the country, jasmine is grown for commercial use in perfumes and Ayurvedic medicines. Considered pure and auspicious, the flowers are also used in religious ceremonies and as offerings to the gods. Kalidasa in his play *Abhijnanasakuntalam* compares Shakuntala to a newly opened jasmine flower.[3]

As I write this, I am made aware of the astonishing variety of fragrant plants available to us in the city. In our small space, in the middle of a bustling metropolis, we have five types of jasmines: star, common, mogra, kamini and *raat ki rani*. We also have two trees of harshingar, champas in three colours, a mature gardenia bush, a gorgeous swarna champa and just outside our house, trees of Persian lilac and alstonia or saptaparni, all adding their special redolence to our lives.

20th

The copse of kosam trees at the end of our lane is in its second flush of red leaves, the crimson a welcome burst of colour in a sea of green. The first and more prominent flush of red occurs in March–April when the tree stands out like a beacon, shining from afar. The second rainy season phase is more subdued, but still striking, and once you start noticing the tree, you seem to see it everywhere. The 2–3 cm long ovoid fruit ripens during the rainy season as well and looks rather pretty hanging in clusters.

The kosam tree is found throughout India, its range extending further to East Asia, Myanmar, Thailand, Malaysia and Indonesia. It is the host tree to the lac insect and produces excellent quality shellac. Oil extracted from its seeds is used to treat a number of ailments, including rheumatism. 'In earlier times, it [the oil] was used to make a hairdressing called "macassar oil" (hence "anti-macassars" on Victorian sofas).'[4]

24th

A beautiful Guru Purnima full moon. This is also called Ashadha Purnima (after the Hindu calendar month) and Vyasa Purnima, being the birth anniversary of Ved Vyas, the author of the Mahabharata. Guru Purnima is the day for worshipping and giving thanks to one's guru, usually a spiritual teacher. Buddhists celebrate Guru Purnima as the day when the Buddha gave his first sermon at Sarnath, after which the Sangh or community of Buddhist monks was formed. In the West this is the buck moon, so called because the antlers of male deer or bucks are fully grown by this time.

This is no special moon; it is neither super, nor blue nor pink, but it is gorgeous all the same. It is also especially luminous, being in perfect alignment with the Earth and the sun because of which sunlight falls fully on the side of the moon facing us on Earth. For me, however, the moon is always fascinating; the magic of a round shiny object in the sky, and all the secrets it holds, is an enduring one.

26th

Was woken up very early in the morning by the sound of rain lashing against the windowpane. The dark of the predawn, coupled with the cocooning silence in the room, lured me into a state of homey restfulness, and my mind drifted back to the various songs and moods of the rain I loved.

> 'On palm-trees shrill,
> On thickets still,

> *On boulders dashing,*
> *On waters splashing,*
> *Like a lute that, smitten, sings,*
> *The rainy music rings.'*
>
> —SUDRAKA, *Mricchakatika*

A tropical downpour drumming on the roof like an impassioned percussionist; the pitter-patter of gentle precipitation on the leaves of the gardenia outside my window; the liquid sunshine of a sunny day shower, with each drop holding a rainbow in its heart; sudden summer squalls that pound the roads for a few violent minutes; rainwater splashing down drain pipes and creating runnels in the flower beds; the determined concerto of a monsoon drencher; wind-driven spindrift dewing my hair as I sit in the veranda; the soft murmur of a drizzle at dusk, which D.H. Lawrence has expressed so eloquently in *Lady Chatterley's Lover*, 'The drizzle of rain was like a veil over the world, mysterious, hushed, not cold.'[5]

Since rainfall in Delhi is largely confined to just three months—July, August and September—the wait for a proper downpour is a long one and the drought, both metaphorically and literally, can last for months at a time. Western disturbances do sometimes bring winter rain, but this is uncertain and often scanty.

27th

Saw a beautiful flowering dwarf bauhinia at a friend's house. Also called white kanchan, the shrub has striking bi-lobed leaves and fragrant white flowers with pretty yellow-tipped stamens. Cultivated mainly as an ornamental plant, though its leaves and bark are sometimes used to treat asthma, it deserves more popularity than it currently enjoys in Delhi.

28th

Heard my first pied cuckoo of the season today, the distinctive 'piyu-piyu' stopping me in my tracks. The bird is a summer visitor to North India from Africa, arriving in May–June, but it is only in the rainy season that it takes off in all directions and becomes visible in our little corner.

The pied cuckoo or chataka (Sanskrit) has been much discussed in ancient Hindu mythology where it has come to symbolize an intense longing. The bird is said to long for the rain, even developing an additional beak on its head (which is its crest) to catch the first drops that fall. Since it is said to drink only rainwater straight from the skies, it keeps up its yearning call all day, begging the clouds to quench its thirst. Adi Shankara, the great mystic of the eighth century, writes in his *Shivananda Lahari*[6]:

> Lord of Gauri! As the swan loves the lotus bed, the Chataka bird the dark cloud, the Koka bird the sun every day and the Chakora bird, the moon—even so, O Lord of beings, my mind desires your lotus feet...

Kalidasa, in his *Meghadutam*, also refers to the chataka, where he considers the sighting of one to be a good omen[7]:

> 'While a friendly breeze impels you gently
> as you loiter along, and here on your left
> the cataka in its pride sings sweetly...'

30th

I have a new friend! For the past couple of weeks, every morning at 6.30, just as my husband and I take our tea out onto the veranda, a house crow flies in and perches on a low-slung branch of the saptaparni growing just across the garden from us. I recognize him now and notice that he likes to announce his presence with a loud cawing. Initially, when I offered him crumbs from my wholewheat biscuit, he was wary and after much procrastination and hopping from branch to wall and back again, he would scoop up the treat immediately and decamp with it. Today, however, he flew down as soon as I offered him the biscuit and ate it whilst standing on the lawn itself, and then looked expectantly at us for another titbit. Tomorrow I intend to keep a piece of roti for him. Might as well feed him healthy stuff!

And then there are the babblers. Moving in a large group consisting of friends and relatives, they descend on the garden in a boisterous gang—hopping around perkily, poking around in the grass, looking under leaves for insects—all the time connected to the others through non-stop chattering. Yesterday morning, we were treated to a very heart-warming display. As soon as I threw a piece of walnut towards the birds, one of them rushed to pick it up. I have observed this about babblers, that once a bird has emerged victorious, the others leave him alone to enjoy the spoils, never chasing or heckling him for a share. The bird with the walnut then hopped across to the flower bed along one side, where Junior was waiting amidst the lantanas and the hibiscus. The parent painstakingly broke the walnut into small pieces before feeding it to the baby, bite by bite, although I personally thought that the youngster looked perfectly capable of foraging for his own breakfast!

We have a resident magpie-robin who likes to take a dip in the bird bath as soon as the water is changed in the morning, and on particularly hot afternoons a treepie also comes to bathe, its body barely fitting into the terracotta vessel. A neem tree just outside the house is a favourite perch for a pair of large green barbets, although I am noticing that their calls have reduced somewhat. Not the koel's though; several times during the day, I hear snatches of his song as he flits from tree to tree.

August

> 'With streaming clouds trumpeting like haughty tuskers,
> with lightning-banners and drum beats of thunder claps,
> in towering majesty, the season of rains
> welcome to lovers, now comes like a king, my love.'
>
> —KALIDASA, *Rtusamharam*, Canto II: Rains

1st

The spring in Delhi can be said to be balmy, the summer scorching and the winter cold and foggy. But the monsoon stubbornly defies any attempt to describe it in such banal terms; it is intense and sensual and more than just the rainy season. It is the daily tracking of the advancing monsoon arc on television and in the newspaper, the excitement when dark clouds finally start to gather on the horizon, the touch and feel of the rain on outstretched hands, hearing the *badra garjat ghanan ghanan*, thunder and lightning on dark rainy nights, the *saundhi mitti ki khushboo* (petrichor) that arises when the first raindrops fall on the parched earth. It is also the irritation when clothes refuse to dry, the overpowering humidity of rainless days and the frustration at the almost daily traffic snarls.

August in Delhi is a sweltering month, with an average high of 36°C and an average low of 29°C. The heat index or felt temperature can reach 47°C, and coupled with the fact that this is the most humid month of the year, the days can become very uncomfortable. It rains for an average of 17 days and the average length of day is just above 13 hours, decreasing from the start to the end of the month by 44 minutes. The chances of a day being muggy remain consistently above 90 per cent throughout the month and on 16 August, which is the muggiest day of the year in Delhi, this number increases to 99 per cent.

2nd

However uncomfortable the weather might become, this is the month of Bhadrapada in the Hindu calendar, a time for clouds and rain. The sensory magic of the monsoon has long been celebrated in Indian mythology and culture; in art, music, poetry, food, cinema, literature.

The Varsha ritu or sawan has inspired an entire range of miniature paintings in a series called Ragamala, which are illustrative paintings of the 16th and 17th centuries depicting on paper various musical ragas. It is a beautiful concept and ascribes to each raga a colour and mood, season, time of day when it is to be sung and its own hero and heroine. The monsoon season is associated with Raga Megha, and most of the Ragamala paintings in this series feature rain-laden clouds, stormy nights, thunder and lightning and lovers enjoying the season or suffering pangs of separation.[1]

The monsoon season has also closely influenced classical, folk and contemporary music in all parts of the country, with each region having evolved its own rhythm and metaphors. Miyan Tansen, the court musician of Akbar, is associated with the famous monsoon raga he authored—'Miyan Ki Malhar'. Raga Malhar has more than 30 ragas under its umbrella, with different lyrical compositions celebrating different aspects of the rain. Laura Leante says of Raga Megha that it 'elicits images of cloudy skies, and rumbling, roaring sounds announcing the rain'.[2] The rains have also inspired a vast body of folk music; kajari and thumri of UP and Bihar, the teej songs of Punjab, bhaleri and shetkari folk music of Maharashtra, among several others. In popular Bollywood monsoon music, songs of the rainy season sing of eroticism, of love and longing, separation and then the coming together of lovers, of frolicking under an umbrella or in wet clothes that cling to the actors' bodies.[3]

Meghadutam by Kalidasa remains the most famous literary work inspired by the monsoons. The poem tells the story of a Yaksha, a celestial being, who, being banished from the kingdom and pining for his wife, tries to send her a message through a rain cloud![4]

> '*As you approach the noble mountain Citrakuta,*
> *he will greet you, O travel-weary Rain-Giver,*
> *and bear you on his head held high: you too*
> *with sharp showers will quench summer's cruel fires.*'
>
> —KALIDASA, Meghadutam

Over the centuries, numerous writers and poets have produced a huge volume of work built around this season of plenty. Amir Khusrau (AD 1325), an Indo–Persian court poet, invokes the rain in many of his ghazals[5]:

> '*The cloud rains as I leave my beloved.*
> *What can a heart do, parting from its love on a day like this.*'

More recently, Rabindranath Tagore celebrates the season in numerous poems, especially in *Abar Esheche Asharh*.[6]

> '*The darkly veiled June has come once again*
> *redolent of the rain-soaked earth;*
> *my heart that had grown weary and old*
> *answers to the call of the marching clouds…*'

(The month of Ashadha lasts from mid-June to mid-July and signals the start of the monsoons.)

4th

August in Delhi means beautiful skies. Clouds paint the vault of heaven blue, grey and white, their magic brush dipped into pots of pastel colours. Some days are a clear cloudless azure, the sunlight then sharp and piercing, while others are mottled, marbled, stippled. A thousand shades of grey, all piled up in a chaotic heap. I have always felt that dark pewter is a colour that offsets beautifully every other shade; yellow laburnum flowers or the red of the gulmohur or even the green of the leaves are all accentuated against a darkening sky.

The importance of monsoons in Indian culture and life can be gauged from the fact that there are 15 or more synonyms for clouds in Sanskrit, and nearly as many for lightning. Since the very dawn of Indian literature, clouds have inspired poets and writers alike. In the *Rigveda* (1500–1000 BCE), the rain cloud Parjanya is given the status of a god alongside the other forces of nature.[7] Clouds have been celebrated as shape-shifting magicians offering respite from scorching days and muggy nights. The colours they assume have been compared to mundane and fantastical objects alike; kajal, indigo robes, Lord Krishna, a wet buffalo's belly, among others!

5th

The second most noticeable thing about August is the luxuriance of growth in the plant kingdom; every leaf, every shrub is renewed, possibly even more than in spring. It is as if all life rejoices in this season of abundance; the parched earth sends out new shoots of greenery, peacocks dance with joy and I can hear frogs croaking in small puddles that build up in the muddy lane behind our house. Flowers stand out here and there in this astonishing verdure sea; the last of the summer blooms, the laburnums and gulmohurs, the fresh white of champas and the pink and purple of the jarul.

The garden is becoming increasingly self-willed in this season of unfettered growth; shrubs and vines refusing to be confined to space that I or the gardener, who comes once a week, have allotted to them. But I like it this way; who am I to apportion pieces of land to my co-inhabitants of this planet that we all share? So if the lantana has spread onto the grass, let it. Nature needs to do as she pleases, especially in a city like Delhi, located in an area of low rainfall and prone to air pollution.

8th

The lushness of greenery forms a pretty backdrop to the lagerstroemias that enjoy a second flush of flowers this month. Delhi has five species of this genus, the most common of which are the jarul and crêpe myrtle, the former as a tree and the latter as a shrub.

Also known as the 'Pride of India' and 'Queen's crêpe myrtle', the jarul is one of the prettiest trees in the city, striking throughout the summer and monsoons. It has big elliptical leaves that turn mahogany-red before falling off; the flowers come in three colours—pink, lilac and white—and are held in big, upright, branching clusters with the older, paler blossoms at the bottom and the newer blooms at the top, giving to the tree a distinctly ombré look. The petals are crinkly and bear a striking resemblance to the crêpe paper after which they are named! The fruit is very interesting too; small and round with a spike at the top, changing in colour from olive to a rich nutty brown with time. When ripe, it splits into a woody flower with five or six segments, each with seeds inside, and makes for a very pretty autumn decoration. E. Blatter and

Walter S. Millard in their book, *Some Beautiful Indian Trees* call jarul the 'Queen of Flowers'[8]:

> The young leaves come out with the blossoms in May. Then the tree covered with great clusters of large mauve flowers is a delight to the eye. Its massed flowers have not the aggressive beauty of the Gul Mohur or the Flame of the Forest but their soft pastel colouring is tenderly attractive and pleasing. Each cluster or panicle of flowers may be quite a foot in length springing from the branch as an upstanding spike, massed with flowers at its base and bearing numerous downy pink and green buds towards its tip.

Jarul is the state flower of Maharashtra and a stamp was issued in its honour in 1993 by the Government of India. Tea made from dried leaves of the jarul has traditionally been used to treat diabetes, and its timber was used to make boats and canoes. Also, it has a very dense and spreading root system, which makes it very effective in areas that need reforestation and erosion control.

Unlike the jarul, the crêpe myrtle is grown mostly as a shrub; its popularity is largely because of its long flowering period of nearly four months. It is a hardy, low-maintenance plant with multiple trunks and a flat top, giving it a distinct bush-shape. It is ubiquitous, growing everywhere—road dividers, small gardens, parks, even neglected verges. But come the flowering season, from May right through the rains, and it is smothered by large, showy clusters of blooms in a range of eye-catching colours—pink, purple, white, crimson, all with the same crinkly texture as the flowers of the jarul.

10th

It has been raining for the past two days, a gentle rim jhim percussing rhythmically on the leaves outside my window. Coming back home in the evening yesterday, I was caught in a traffic snarl, which is a common occurrence in Delhi during the monsoons. Fortunately we were stuck right next to a flowering katsagon and I spent a very pleasant 10 minutes admiring the conspicuous flowers and fruit pods that appear almost simultaneously during the rains. An interesting aside—the flowers, fruits and leaves of this tree are all woolly and fuzzy. The flowers are big and showy, in a pretty shade of lemon yellow, but are somewhat overshadowed by the long, curly, cylindrical fruit pods that cluster at the end of the branches and which remind me of the pencil-shaped erasers the children had as kids, which you could twist and curl as you pleased. The flowers are similar to those of the sausage tree and like the latter, open at night and are pollinated by fruit bats. The leaves are large and broad, with the undersides covered in reddish fuzz.

14th

There is always one day towards the middle of August, when during a break in the rains, on a sunny afternoon, I sense a shift in the light and know that the seasons are changing. The change is gradual, imperceptible even, but inexorable. The days when it doesn't rain are still hot and humid and uncomfortable, but the dappled shadows cast by the trees near our house are different now, and the sunlight, if not mellow, is not harsh either. The promise of winter is carried on the wind and I look forward to the cooler, shorter days.

The sausage tree is hung with both fruit and flower and makes for a very pretty sight. The weeping bottlebrush also bears both fruit and flower together this month, as do most of the jaruls. The red cassia, also called the Ceylon senna, is in bloom, and a tree near our house catches my attention whenever I pass that way. Each drooping branch is covered in clusters of terracotta red flowers, which look lovely against the feather-compound leaves. Old blooms change in hue from red to dark pink. The bark of this tree is used to produce an oil, which has medicinal value.

The copperpod (also known as peeli gulmohur) is in its second flush of flowers, the tip of each branch as if dipped in bright yellow paint, imbuing the entire canopy with a golden glow. The karanj is also in its second phase of flowering, the first occurring in April–May. There is an entire row of these trees just outside our colony. Another batch of tiny yellow figs on the laurel fig. A handsome tree in all seasons with a huge spreading canopy and aerial roots that do not quite reach the ground, stopping midway to wrap themselves around the trunk. The goolar by the side of our house is still fruiting, the green and red and orange figs, representing different stages of ripeness, always make me think of traffic lights! Perfectly round figs, laid out along the trunk, like a table heaped for community feasting! And after a night of revelry by the bats, the driveway looks as if something has exploded on it!

18th

Every season paints the landscape in its own special hues, and at no time is this more apparent than during the monsoon. August in Delhi is a complex synthesis of grey and blue and green. Delhi's normally wishy-washy sky now takes on more character; the colours are bold, intense even; clouds of pewter, slate and gunmetal crowd the horizon, and on a clear day, the bright azure of a sky observed through a squeaky clean, washed atmosphere dazzles, as do the saturated greens of lush verdant growth.

Just behind our house is a handsome chikrassy tree, also called Indian redwood and the East Indian mahogany. This imposing tree has a spreading canopy and distinctive drooping branches. It appears to be in new leaf this month (probably the second flush?) and the young leaflets at the end of each feather-compound leaf are a pretty shade of red. The leaves are very interesting, being totally asymmetrical at the base.

The earpod wattle is in bloom, and the long baton-shaped flower clusters are a striking sunshine yellow. Everything about this interesting tree is curly-wurly; the trunk is crooked, the leaves are long, leathery and curved and the fruit pods are coiled. The earpod wattle is an evergreen tree native to northern Australia and can withstand dry drought-like conditions. Its wood is used mostly as a fuel, and gum extracted from the trunk has some commercial value.

The yellow oleander, which also blooms during the rainy season, usually has eye-catching, butter-coloured, trumpet-shaped flowers, although I find that the white flowering one growing in our back garden is prettier. The blooms twinkle like stars in the gathering dusk every evening! The leaves are long and narrow and arranged in spirals. Locally known as kaner, this small tree, which I have often seen growing as a shrub, is fairly common in parks and along roads. Recently, at a local nursery, I saw pink and magenta and orange flowered
kaners, which the gardener told me were new hybrid varieties.

20th

Small green fruit of the Persian lilac tree on the driveway in the morning; all the berries were perfectly halved, looking on the inside exactly like miniature apples! Either the fruit splits on impact or the parrots have been at it.

No rain for the past week or so, and the days are clammy and dense. Today, however, a light wind picked up in the late afternoon, rustling in the topmost branches of the neem and Persian lilac, and by evening it had herded in stratocumulus, which is the most frequently occurring monsoon cloud type. (Other common monsoon clouds are altostratus and cirrostratus.)

August in Delhi is dynamic, the interplay of sun and shade, light and dark. Everything is in constant motion; ephemeral and unfixed. Running sky, dancing raindrops, rushing wind, rolling thunder, scudding clouds, rustling leaves, gliding eagles, hovering dragonflies, fluttering butterflies; the eternal flow of the universe!

22nd

A bewitching full moon, in conjunction with Jupiter, in the constellation Capricorn, with the planet being visible just above our satellite. Another planet, the extra bright 'evening star', Venus, has been visible since late evening, even before the sun had set fully, making me want to linger outside a while longer. Tonight's full moon is also a blue moon, in the sense that it is the third full moon of four this summer season (we follow the dates of the official summer season of the West). Usually every season has three full moons, but if four occur then the third is called a blue moon. The second full moon in one month is also called a blue moon.

The August full moon is known as Shravan Purnima and is considered to be a very auspicious day in the Hindu calendar; the festival of Raksha Bandhan or Rakhi, which celebrates the bond between brothers and sisters, is celebrated on this day. In Western etymology, this is the sturgeon moon of the US and the grain moon of the UK, so named because it marks the beginning of the grain harvest.

25th

A parliament of crows has convened on one of the smaller goolars outside our house, clothing its topmost branches in regimental black, and the afternoon is suddenly filled with cacophonous squawking. As is often the case in a democracy, the seniors are unable to ensure a dignified silence! A pair sometimes nests in the tree and I feel that the juvenile who has been visiting the garden with his parents has his nursery on one of the topmost branches.

In the late afternoon, a white-throated kingfisher, with his unmistakable huge red beak, perched on the electric pole right outside our house. A common resident of Delhi, it is often seen on wires and exposed branches of trees, even far from water, although I think that the largish puddle, formed due to a leaking garden tap, right under the pole is what attracts him to this particular place. I have been hearing his distinctive laughing call, off and on throughout the month, usually in the early mornings, and have often rushed outside to catch a glimpse of brilliant blue as he flies across!

29th

Just behind our row of houses is a narrow strip of government land, fenced along the open side, and mostly left to its own devices. The gardeners put in an occasional appearance, armed with big secateurs and other tools of their trade, but I have never noticed any visible pruning and cleaning, which is just as well because I like it to remain in its native overgrown state. It is used once in a while by the more adventurous and intrepid walkers amongst us.

Today early morning, much against my wishes, I was bullied into taking a saunter down the unkempt path which, after the recent rains, has transformed into one big puddle.

But was I glad I went!

On the road in front of us was a family of common babblers making their usual racket. But they had a juvenile pied cuckoo in tow, and it was amusing to see the babbler parents taking turns to feed a bird bigger than themselves! Pied crested cuckoos are brood-parasitic and leave their eggs most commonly in the nests of the turdoides babblers. In this case the two juveniles, the babbler and the cuckoo, looked old enough to hunt for food themselves, although the cuckoo seemed to be the more efficient of the two, pulling out an insect from the undergrowth on more than one occasion.

I have often wondered why the parent birds are not able to distinguish between their own chicks and the interloper. After all, birds are intelligent beings able to find their way halfway across the world, mark out their territories, woo their mates—often with elaborate and complicated dance steps—and even fight for their rights if necessary. How is it then that they are unable to intuit that the cuckoo or the koel is not one of them? Is it because the parents grow to love the babies they bring up, or because they parent on autopilot, simply feeding the hungry mouths open in front of them?

Whatever the reason, it was rewarding to witness the inscrutable workings of nature from so close up!

September

*'The earth is bright with Kasa blossoms,
nights with the cool rays of the moon;
streams are lively with flocks of wild geese
and pool are strewn with lotuses;
groves are lovely with flower-laden trees
and gardens white with fragrant jasmines.'*

—KALIDASA, *Rtusamharam, Canto III: Autumn*

1st

September is a time for retreating monsoons, clear days and bright sunshine, especially towards the second half of the month. It is the start of autumn or Sharad ritu, coinciding with the month of Bhadra and Ashwin in the Hindu calendar. The latter is the seventh month of the Hindu calendar and technically begins on the new moon after the autumnal equinox. An interesting fact; Ashwin is also the name of the first star that appears in the evening sky and also refers to the glow of sunrise and sunset.

September is the start of the autumn festivals and the Sharad Navratras usually fall within this month. The rains end by the middle of the month, leaving behind cooler and pleasanter weather. The average temperature hovers between

35°C and 25°C with an average relative humidity of above 50 per cent. This translates into an average of 10 rainy days in the month. Bright sunny days are still uncomfortably hot, but the very feeling that we are into autumn and can look forward to cooler days ahead is like a blast of fresh air!

With the breeding and nesting season having by and large ended, the birds are about their normal business again, with some having gone into moult. Trees are clothed in monsoon verdure and many are flowering.

2nd

The start of September. I sit on the veranda in the early morning, immersed in the soundscape; thunder rolling in the distance and then the crackle of lightning. It is overcast to the point of being dark and absolutely still. The lull before the storm; even the birds are quiet. Then the rain starts, falling in sheets, and I try to identify the different notes in its music; the loud drumming on the stone-paved driveway, the pitter-patter of the raindrops on the leaves of the nag champa growing right next to where I am sitting; the clatter of the water as it rushes down the drainpipe on the roof. And then as suddenly as it began, the music fades to a whisper, the last notes ebbing away, carried on the wind. Almost immediately the birds are out, searching for insects flooded out of their homes. The melodious whistling of the magpie-robin, the cawing of my crow, the casual conversation between our resident mynas, the screeching of parakeets overhead, the cheerful calls of the bulbuls, the sudden alarm call of the squirrels on seeing a cat. These are the soundtracks of life, the melodies that are ever present for those who have ears to hear them!

3rd

Yesterday was the September day with the most rain in nearly two decades. It was raining today morning too, not torrential like yesterday, but a steady *sawan ki jhadi*. As soon as we took our morning cups of tea outside to the veranda, searching for a dry spot to drag the chairs into, we were greeted by a flock of hungry, bedraggled babblers demanding to be fed. I shared my atta cookies with them, which they gobbled up within minutes.

Some birdsongs have a very strong sense of place and time. The call of the koel invokes dry, scorching summer days while that of the pied cuckoo is evocative of thunder and louring clouds and rain. Nothing so picturesque for the babblers though I'm afraid! Their high-pitched, strident screeching always tells me that they've come looking for food!

4th

For the past 10 days or so, I have noticed that late evenings are suffused with a strange light; just before sundown an orange-pinkish glow fills the western sky and bounces off every surface. The yellow flowers of the allamanda and the green bamboo fence to the front of the garden turn incandescent for a brief moment of glory before the growing dusk swallows them up! The most plausible reason for this colourful sunset is the presence of mid- and high-level clouds that capture the sunlight and reflect it to the ground. High-level cirrus clouds create an orange glow while altocumulus clouds are usually responsible for more peaches and pinks in the sky.

I came across a small colony of baya weaver birds at the most unlikely of places—in a corner of a small neglected park in one of the old housing colonies of Gurugram. The striking birds and their even more unmistakable nests—'a swinging retort-shaped structure with long vertical entrance tube, compactly woven out of strips of paddy leaf and rough-edged grasses, suspended in clusters from twigs usually over water'—immediately caught my attention.[1] There seemed to be no water body around and I wondered where the birds procured the nesting material from.

6th

The garden is doing surprisingly well for this time of year;

the purple gomphrenas are still blooming profusely—the white and pink ones didn't last beyond a month—and the periwinkles, while not at their most prolific, are holding on, adding colour to the beds from where the portulacas have died down. The moss roses are blooming too and are a magnet for the bees. The rudraksh tree I planted last year has grown well, but I am very doubtful about it bearing any nuts in our Delhi weather.

The butterfly pea or aparajita vine is flowering, but unlike most other creepers, the foliage—leaves and flowers—are only present in the top one-third of the plant, with the rest being the woody stalks that have wrapped around themselves with much dexterity, giving the plant a nice strong base. It is a prolific spreader, with new saplings sprouting almost daily, cobalt blue flowers appearing suddenly in the midst of the yellow lantana shrubs growing underneath! The seed pods have remained on the creeper ever since the last flowering finished so the plant has both flowers and pods now. The flowers are edible and used in many South Asian cuisines. They also make a delightful tea, rich blue in colour, which changes to a pinkish-purple hue with the addition of lemon and honey. The plant has long been used in Ayurvedic and traditional Chinese medicine.[2]

9th

A lazy sort of morning made lazier by the presence of four plump, yellow-footed green pigeons on the goolar by the side of the house. Frequent observation has convinced me that these birds are the couch potatoes of the avian world, sitting motionless on the same branch for long stretches of time until struck by a pang of hunger, when they might jump onto an adjoining branch or simply bend over, stretching at impossible angles to get to any straggler berries. Their colouration provides them with an excellent camouflage and it is only movement that gives away their presence. So maybe they have a good reason for their sloth. E.H.A. in his delightful book informs us that these birds, despite being fruit eaters, are not very active on banyan trees because 'the Green Pigeon cannot dig holes in fruits: it swallows them whole. Now the Banian fig is tough and so firmly joined to the twig that the Green Pigeon has not strength to pull it off.'[3]

The general lassitude of these pigeons extends to the hunt for twigs during the nesting season as well. I have noticed that the bird will come to the same tree again and again in its search for suitable material; not for it to get out of its comfort zone! Once it has alighted on its chosen branch, it will attempt to get to the sticks in its surroundings with as little movement as possible, only occasionally flying to a different tree.

12th

The butterflies are back! And so are the birds at the feeders, now that the frenzy of finding partners and raising a family is over. Our resident myna pair has reduced its bellicose swagger somewhat and we often have a small flock of their young and friends in the garden these days, especially in the early evening when the grass is usually watered. Salim Ali says of the extended myna family in September, 'It is fun, watching these family parties of Mynas at work on a wet grass field. They hop along swiftly and jauntily after the grasshoppers, and often jump up and chase the insects in the air for a short distance. There is no escape for the enemy: if it succeeds in dodging one bird it is sure to be snatched up by the next.'[4]

The magpie-robins are accompanied by juniors, who keep up a constant 'chur-chur' throughout the day. Could it be their sub-song? There is a palpable decrease in aggression and excitement in the avian world, and the red-vented bulbuls no longer create a furore at the appearance of the treepie when he visits for tasty titbits! Today morning, he was joined by a beautiful male rose-ringed parakeet at the small feeding table. Complete bonhomie! The purple sunbirds are enjoying the calliandra and allamanda flowers, and yesterday I saw that the male of our pair is back in his eclipse plumage. The pigeons are a menace as always, threatening the silverbills at the feeders and insisting on perching on pots too small for them. I wouldn't mind their presence if they learnt to exist in harmony with the smaller birds.

16th

On Lodhi Road, just a little way behind the main row of buildings in a quiet cul-de-sac, saw a largish group of black kites roosting in a grove of ancient trees. A beautiful late afternoon, rainy and overcast, gave way to a spectacular sunset against which the silhouettes of the big birds stood out in stark contrast.

Saw an unfamiliar tree with pretty yellow flowers and had to come home and check it out from the photograph I took. It turned out to be the Indian tulip or bhendi tree, not very common in Delhi because of its preference for a more humid climate. However, it seems to have several medicinal uses; its bark being used as a diuretic and to treat indigestion and cough.

17th

A blustery day, a blustery night and now another blustery morning; dark and overcast with flocks of dragonflies flying low. Dragonflies have the most interesting nicknames of all, which include 'devil's darning needle', 'snake doctor' and 'sewing needle', among others.[5]

A flock, or should I say, a pandemonium of alexandrine parakeets on the peepal, shrieking and calling almost non-stop. When they leave there is a sudden silence and only gradually does the cheerful chirping of the bulbuls become audible. Two large brown-headed barbets are regulars at the same tree, and they too are unusually vocal today. Maybe the parakeets have inspired them! To add to the cacophonous babel, a black-rumped flameback woodpecker flies in, his distinctive call—'kee-kee-kee'—announcing his presence even before I spot him.

The terrace affords me ringside views of the semicircular amphitheatre created by mature trees just outside our gate, where every day troupe after troupe puts up its theatrical performance; humorous or romantic or, at times, even a crime thriller. Rose-ringed parakeets amuse with their acrobatics while the mynas are willing to fight to the end for nest holes. Barbets call, announcing the onset of summer while drongos herald the beginning of autumn. It is an ever-changing tableau; a record of the changing seasons; leaves drift down, new buds appear and seed pods mature at their appointed hour. 'To everything there is season, and time to every purpose under the heaven: A time to be born, and a time to die; a time to plant, and a time to pluck up that which is planted.'[6]

Trees are like books, full of stories and myths and strange happenings. This little corner, so insignificant in a large city constantly on the move, is full of anecdotes and narratives; secrets which it is willing to share with whoever has the time to pause and listen. But how many of us do, caught up as we are in the daily grind of life? Ruskin Bond certainly does and shares his list of favourite trees.[7]

> The banyan, with its great branches spreading to form roots and intricate passageways. The peepul, with its beautiful heart-shaped leaf catching the breeze and fluttering even on the stillest of days. It is always cool under the peepul. The jacaranda and the gulmohur bursting into blossom with the coming of summer. The cherries, peaches and apricots flowering in the hills; the tall handsome chestnuts and the whispering deodars.

18th

The full moon is still two days away, but the night sky is full of wonder. The moon adds to a large part of the magic, of course, but it is the planets—Venus, Jupiter and Saturn—that contribute to the wow factor. The moon passes only a few degrees north of Jupiter and Saturn and forms a rough arc with the two gas giants in the eastern sky. Venus, the evening star, on the other hand, is best viewed in the west immediately after sunset, where, although it rides low in the sky, its brightness makes it unmistakable.

Then there is the Summer Triangle—the asterism, a pattern made by bright stars from different constellations—formed by Vega, Deneb and Altair. Though visible throughout the year in the northern hemisphere, it is best observed during the summer. In Delhi, however, where the summer sky is often blighted by heat and dust, the most conducive time to see it is in September, when the monsoon has rid the atmosphere of all obstructing pollutants. Then, on a clear, cloudless evening, look overhead in the gathering dusk and you can see it clearly—the triangle in the heavens! I have been fascinated with this starlit geometry ever since I was a child and my mother showed me a photograph of the Milky Way flowing like a diamond-studded river between the stars Vega and Altair. That feeling of enchantment has stayed with me to this day!

19th

A moody day, tumultuous dark clouds piled up in a multi-layered expanse, the roads carpeted with dried leaves, a soft drizzle. The afternoon grew dark and the greens of the trees were illuminated in a strange light, making my pulse quicken in anticipation. And then the rain started pouring down in sheets, the myriad hues of grey painting the sky pewter, steel and dove grey, transforming it into a uniform opalescent moonstone. Almost simultaneously the cloud cover changed into nimbostratus, a light grey featureless cloud sheet that almost always brings rain.

> Surreptitiously and without fanfare is how the Nimbostratus arrives. It generally results from the thickening and lowering of Altostratus. Since one cloud leads to the other, the point of distinction between the Alto and Nimbostratus is rather academic.[8]

20th

A beautiful full moon dominates the horizon. This is the harvest moon of the west, so named because of the moon rising shortly after sunset at this time of the year, resulting in an exceptionally bright late evening, which traditionally helped farmers to harvest the summer crop. This is the bhadrapada moon of the Hindus and marks the start of the 'Pitru Paksha', the fortnight in which ancestors are revered and offered food. The Buddhists celebrate this day as the Madhu Purnima.

The autumnal equinox is three days away and falls on the 23rd. This is the day that the sun crosses the celestial equator going south, and along the way promises us cooler days and pleasanter times. On this day the Earth's axis is tilted neither away nor towards the sun, leading to days and nights of equal duration all over the world. It is the official start of autumn in the northern hemisphere and is the Sharad ritu of the Hindu calendar. The sun will start rising more from the south each day and will stop shining directly in my face while I have my morning tea in the veranda outside. An interesting fun fact; the autumnal equinox, like the spring one, shifts through the different constellations over time. The former passed from the constellation Libra to Virgo in the year 730 BC while the latter will move into Capricorn in the year 4312![9] Never a dull moment in the cosmos!

26th

A row of floss-silk trees some distance down our road has put on a very impressive show. Clothed in beautiful pink flowers, the largely leafless trees with their green barks are almost flamboyant in appearance. The flowers appear in September and October, and although they are undeniably lovely, the tree holds its own for the rest of the year as well because of its very interesting trunk that reminds me of the 'green man' illustrations in children's books. The trunk is a pretty shade of green in young trees and is studded with conical prickles that fall off with time. 'Bark, a startling grass-green at first. As it stretches over the widening trunk, the green patches become furrows, separated by ridges of newer grey bark.'[10]

There is a large ronjh tree right outside our gate, whose trunk has turned dark with age. The gardener tells me that this is a very old tree, from long before houses were constructed in the area, and that is probably why it has strange black growth all over the stems. Ronjh belongs to the thorny mimosa family and is native to Delhi, superbly adapted to its semi-arid climate. This tenacious tree is drought, fire and frost resistant, and manages to survive rather well even in stony, dry, sandy soil. Every evening, after the sun has set, the leaves close, much like those of the calliandra.

Ronjh is in full bloom this month, the tiny, light-yellow flowers borne on the edges of branches enveloping the tree in a golden haze. The flowers are fuzzy and arranged in round clusters, giving the impression of fluff balls. This entire month the entrance to our house seems laid out in a furry carpet, almost too delicate to step on!

The bark of the tree is used in traditional medicine, but has a more interesting purpose; to distil liquor, giving it its alternate name of *sharab ki keekar*! In the drier reaches of its range, the tree provides very useful fodder during the dry summer months.

28th

Saw the rain travel for the first time in all my life. It had been overcast since the morning; dark and still with distant thunder occasionally rending the air. Deciding to go for a stroll, I was stopped in my tracks by a loud booming sound. Looking up I saw, to my amazement, a sudden squall two houses down from us. A second later the downpour had travelled to where I was standing and the next moment it had crossed over to the neighbouring house. Magic!

Shortly after lunch, saw two pairs of grey hornbills fly onto a small leafless tree a short distance from our house. Together, but at a distance from each other, one pair settled on one of the lower branches, while the other on an adjoining one slightly higher up. And when one bird of the pair hopped onto a different branch, the second one did too, sitting touching each other all the time. Made me think of two couples on a double date!

October

It was October again [...] a glorious October, all red and gold, with mellow mornings when the valleys were filled with delicate mists as if the spirit of autumn had poured them in for the sun to drain—amethyst, purple, silver, rose, and smoke-blue.[1]

2nd

October is the start of the pleasant season in Delhi, even though it continues to be rather hot, especially during midday. The average temperature is in the range between a maximum of 36°C and a minimum of 26°C, with the mean average temperature for the month hovering around 32°C. The length of day is gradually decreasing, with the sunrise and sunset being at 6.13 a.m. and 6.08 p.m., respectively, on the first of the month, and 6.32 a.m. and 5.37 p.m. on the 31st. That gives an average day length of eleven and a half hours, and maybe this change, although tiny, is what gives me hope that cooler times are coming! October is the driest month of the year in Delhi, with the average number of rainy days being just one. It is also very calm most of the time.

4th

The saptaparni is in bloom and the city is awash with fragrance. The ball-like clusters of tiny greenish-white flowers give off a heady scent, especially at dusk, that reaches you even before you spot the tree so that a walk along any saptaparni-lined avenue is a delight for the senses.

A beautiful perfumed night; the intoxicating fragrance of the harshingar mingles with the equally seductive fragrance of the saptaparni or alstonia as soon as I step out of the kitchen door. In the morning the driveway will be carpeted in harshingar blooms, which I gather and bring inside. The lane outside our house is lined on one side (the side nearer to us) by alternating saptaparni and harshingar trees, three of the former, and sandwiched between them, two of the latter. All five are mature trees, smothered in flowers, and as I walk down the road at sundown, the two distinct fragrances, one spicy and the other sweet, interweave and blend by strange sorcery into an aroma so sweet that I try to linger on that otherwise unremarkable stretch of road for as long as I can.

5th

My daughter loves hibiscuses and we have quite a few, planted

in terracotta pots; I find that these do better than the ones we grow in the ground, which become prone to mites more often. The entire summer we have been treated to a profusion of glorious blooms, in hot tropical colours; pinks and reds and yellows and whites and best of all, a late-blooming orange, which we had all but given up on!

Gazing skywards, I make out Jupiter and Saturn in the south-east and two stars of the summer triangle, Vega and Altair, directly overhead. Jupiter, at a magnitude of -2.7 is so bright as to appear almost artificial, and the mind boggles at the thought of light taking 37 minutes to reach us across the nearly 680 million kilometres separating us from the gas giant!

6th

We were out last evening driving on an open stretch of road, from where we had a comparatively unobstructed view of the sky, which is a treat in an uncompromisingly urban landscape. Hundreds of rose-ringed parakeets flying overhead, probably on their way home to their roosts, skein after skein of screeching birds, undulating wave-like. I don't really know where their roosts are, but I have noticed that the birds always fly towards the north at dusk, so they probably have their dorms several kilometres away on the outskirts of the city, maybe near farmlands. Parakeets and, in much smaller numbers, crows are the only birds that I have seen returning to their roosts en masse every evening. And they are noisy travellers, chattering constantly while on the wing!

8th

The weather is changing and early mornings are now cool and breezy. Nearly fifteen parrots at the feeders, keeping up a constant chattering; a casual conversation that ever so often descends into a slanging match. They are the undisputed lords; even the silly pigeons don't mess with them. Only the peacock gets priority.

A pair of yellow-footed green pigeons on the topmost branches of a young peepal, sitting motionless, enjoying the sunshine. A black kite, which had been scouting around for some time, doing his rounds, finally managed to snare some breakfast—which, through the binoculars, seemed to be a small rodent—with which it settled on an exposed perch nearby. However, to my surprise, it was almost immediately besieged by crows, at least five or six of them, who set up an incessant cawing, and if the cacophony was not enough, one of them had the audacity to lunge for the catch. The kite fought them off valiantly for some time, but being vastly outnumbered, ultimately gave up and flew off. The prize, which fell from its claws, was skilfully snatched up mid-flight by a crow, who beat a hasty retreat with it, although considering the number of his brethren who gave him chase, I wonder if he got to savour it at all! M. Krishnan tells of a similar incident between a shikra and a couple of crows, with the shikra fighting the crows courageously and driving them off. He further writes[2]:

> I would much like to tell you how the victor returned to the hard-won meal and consumed it in triumph, but, in fact, this incident ended even more like a story. For, while the hawk was routing its enemies, a third crow made an unobtrusive appearance on the scene, by a rear entrance, and flew away with the dead lizard even more unobtrusively!

And so peace returned to our little corner once again. All through the commotion, the yellow-footed green pigeons continued to bask in the sun, staying put at their place, totally unimpressed by the excitement in their neighbourhood. Only now they had been joined by our resident pair of pied mynas, and sometime later, a pair of black-rumped flameback woodpeckers on the goolar adjoining the aforementioned perch, their red and yellow colouration shining against the pale bark in the morning sunlight.

10th

October has its fair share of flowering trees. The kassod is in bloom, its yellow flowers striking against the glossy leaves. The flowers of the glaucous cassia are a brighter shade of yellow, although the leaves of the two look very similar at a glance. The copperpod is in its second flush and the long upright flower stalks, bearing the pretty bright yellow flowers—which have the texture of crinkly craft paper—are unmistakable. Even so, I think I prefer the nearly indistinguishable African wattle, whose flowers are a more pleasing shade of yellow. The two trees are so alike that they share the same Hindi name—peeli gulmohur.

The kaniar (*Bauhinia purpurea*) is in bloom and the showy pink flowers, offset beautifully by the blue skies of October, enliven many a dull morning. I long confused the kaniar with the kachnar, but have since learnt to tell the two apart by the shape of their leaves, and of course, the flowers. The flowers of the kaniar have longer petals than those of the kachnar and are a more unusual shade of pink. The petals of the former also do not overlap, which is why it is called the butterfly tree. But both are equally beautiful, as are the flowers of the third bauhinia of Delhi, the Hong Kong orchid tree. E. Blatter and Walter S. Millard say of the kaniar, '…large purple, deep-rose to lilac flowers appear amongst the foliage from September to December. The flowers are very fragrant, and are visited by numerous bees, by whose agency pollination is effected.'[3]

Another tree that is arresting when in full bloom is the red cassia, which bears its pinkish-reddish blooms through the rainy season well into October. The jhinjheri is flaunting its striking pods, woody and rather large, which are in complete variance to the unremarkable flowers. The ronjh continues to look like a giant fuzz ball with its tiny creamy white blooms that have completely taken over the otherwise inconspicuous tree. The sausage tree has finished flowering, but is still interesting; the long rope-like stalks now bear large, woody, cucumber-shaped fruits. The pods of the katsagon are longer and greener with a surprising sinuosity.

Just around the bend from where we live is a small grove of young *Tecoma stans* or yellow bells, which, although trees, have a shrub-like appearance. As the name suggests, they bear striking yellow trumpet-shaped flowers which hang in showy clusters at the ends of the branches. The flowering season lasts from summer to late autumn and every time I walk that side, I have the fanciful notion that the blooms have captured the sun in their petals!

13th

Serendipity! At about five in the evening when I went into the garden, wondering whether it was cool enough to have tea outside, I saw a male shikra sitting on the water pot hanging from the small jatropha tree in the corner. He was a beautiful sight; perched nonchalantly on the edge of the terracotta vessel, unconcerned at the goings on around him. Just above him, a pair of crow pheasants frolicked in the calliandra shrubs; they are occasional visitors and were there yesterday too. But to come back to the shikra; he seemed to be a young male and still had some dark markings on his upper parts. After staying nearly motionless for about five minutes, he jumped into the water pot and enjoyed a leisurely bath that I captured on my phone! Soon afterwards he flew away to come back again almost immediately. But not for long; a flock of babblers mobbed him, virtually dive-bombing the poor chap. I wonder what he thinks of his life; unwelcome wherever he goes.

15th

The koel is still the first to sing in the morning, its mellifluous notes ringing out in the semi-dark of the early dawn. Surprisingly, however, it is mostly quiet for the rest of the day. The next to call out are the drongos, followed by the magpie-robins, babblers, bulbuls and parakeets. The koel and drongo are the soloists of this group of yodellers, with the rest making up the chorus.

Over time, these birds have become my friends and I recognize their individual quirks. I know that the crow will eat one piece of roti or biscuit while on the ground, hastily gulping it down, before flying off with another few in his beak. The magpie-robin needs a drink of water after every few bites, while the laughing dove likes to skulk along the very edges of the garden, seldom if ever, coming out into the open. After some time she will fly to the round terracotta bird feeders filled with bajra and jowar and will hop right in, staying there until she has had her fill, with only her neck visible above the rim of the pot. The Eurasian collared dove is bolder and will forage in the centre of the garden with the other birds.

The red-vented bulbul weighs his options from an overhanging branch of the allamanda creeper before venturing down cautiously. His red-whiskered cousin is even less bold and needs considerably more persuasion before alighting on the ground to feed. However, he has mastered the art of balancing on the feeders and visits them for a quick snack throughout the day. He is also adept at quick aerial sallies to get to the feed while the bigger birds are looking the other way. The tailorbird is another regular at the small heart-shaped feeder;

his jaunty ways enliven my day and I look out for him every hour or so.

16th

Our resident myna pair is very tame and will come looking for titbits right into the veranda, where I am enjoying my morning cup of tea. They are, however, deplorably quarrelsome and are forever bullying the smaller birds. The brahminy mynas or starlings, who are occasional visitors to the garden, are not as confident and like to keep their distance. They look so neat and well-groomed, much like the pied mynas who condescend to join the party only once in a while. The grey babblers, on the other hand, descend in raucous hordes, bringing their own brand of excitement to the garden, swinging from feeders that are too small for them, picking out peanuts selectively from the mixture of bird feed and bajra. The group has many youngsters, who although fully capable of feeding themselves, nevertheless follow the parents around noisily. Surprisingly, however, I have often seen three or more adults feeding the same chick, creating a complicated family structure which I don't really understand, but which Malcolm MacDonald has explained rather well.[4]

> Many times […] I watched more than two parents feed the same unfledged family. Several times I saw four adult birds, all with insects in their beaks, queue up on branches beside a nest and await their turn to feed the young. One after another they went to the nest-edge, bent their heads to deposit a viand in some chick's wide-open mouth, and then flew away to make room for the next supplier—like a line of waiters, bringing a succession of dishes to a party of banqueters.

The treepie, when he visits, is very noisy, calling out in his harsh guttural voice all the time. Despite his size, he has become quite adept at picking out his favourite snacks from the mix at the feeders, often having to resort to considerable acrobatics to do so. He also flies down to the garden to pick up titbits of roti or biscuit, although unlike the mynas and the babblers, he doesn't linger.

17th

A pair of rose-ringed parakeets, after having had their fill of seeds and grain every morning, shift to the flowering calliandra shrubs to feed on the flowers, and I don't think I have seen a more adorable sight than a green parakeet with a red pom-pom in his beak! This he holds down with his feet, and plucks out the petals one by one, attempting to get to the nectar at the centre. A pair of purple sunbirds is also active throughout the day on the same bush. The male, in his distinctive eclipse plumage, has made the garden his home and is here morning to evening. He is partial to the large banana flower hanging from the banana plant in the corner, as are the bulbuls who, in an attempt to get to the sweetness within, hover around it sunbird style, wings beating rapidly. Although my daughter insists that the flower has to be cut for the baby bananas to mature fully, she doesn't have the heart to deprive the birds of their little treat, and so there it stays!

No description of the garden birds is complete without a mention of my favourites, the silverbills and the sparrows, who have been on child-rearing furlough throughout the summer and have only last month resurfaced after completing their duties. The silverbills (earlier called white-throated munias) are one of the earliest to get to the feeders in the morning and are never distracted by any titbits on the ground, sticking firmly to their favourite seeds and grains. They form small flocks of up to 20 birds, keeping up constant contact calls, and are very companionable, sharing their space with the sparrows. Their favourites are the bottle feeders, and they will shift to the bowls only when pushed out by the bigger birds.

Our friendly neighbourhood peacock, who appears to be a young male, with his train underdeveloped, is a regular at the terracotta bowl I keep in a corner of the garden just for him, thinking that he won't be able to get to the other feeders. But he's a clever one; craning his neck to reach the tray placed upon a stand, he polishes off everything, every last scrap of grain and seed before venturing boldly to where we're sitting to see if he can snare some little snacks.

19th

Early morning rain brought on by western disturbances originating in the Mediterranean Sea. I love these sudden extratropical storms that break the monotony of a largely dry Delhi winter. Later, the afternoon shines with technicolour brilliance, the morning downpour having polished to high gloss every leaf and blade of grass so that the colours seem almost artificial in the afternoon sun.

> '...there is a harmony
> In autumn, and a lustre in its sky,
> Which through the summer is not heard or seen,
> As if it could not be, as if it had not been!'
>
> —PERCY BYSSHE SHELLEY, Hymn to Intellectual Beauty

So many greens and pinks of the kaniars and the pink cassia. The ixoras are all flowering too; the two pinks, white, red, orange and yellow. The yellow had withered away suddenly a couple of months ago—probably an attack by termites—and I have been able to find a replacement only after considerable searching.

The afternoon sunlight, sharpened by the rain-washed atmosphere, halves every wall, the garden, even the roads, so that you're either fully in it or out of it. Light and dark, with no shades of grey in between! The weather affects my mood intensely, much like music does, and I find that I am happiest on the coldest days with mist and fog swirling in great waves or on dark overcast mornings, heavy with the promise of rain. But then, I love cold, windy, sunny days too, when the wind whips leaves into great eddies. Sadly, we have really few windy days in Delhi, days when the wind sneaks in under your clothes and gets all entangled in your hair. Incidentally, October is the calmest month of the year in Delhi, with an average wind speed of only about 7 km an hour, which the Beaufort scale describes as a light breeze.

20th

I think I finally realize why the dove is the symbol of peace everywhere. Or is it? The pair of laughing doves who frequent our garden and are regulars at the feeders are completely non-violent and will move aside obligingly at the first sign of aggressiveness by the mynas and the babblers. They are unobtrusive and unassuming birds, keeping to the margins of the garden, where they skulk around in the undergrowth, very often feeding on the seeds that fall out of the feeders. But—and this is a very emphatic but—they are very territorial and will chase away their fellow brethren by advancing on them menacingly, head held low and body taut. Our resident laughing dove has perfected this art and is fully capable of chasing away half a dozen interlopers all by himself. Yesterday, I saw him pecking a particularly recalcitrant female viciously on her back, coming away with a soft feather in his beak!

The pair of Eurasian collared doves who frequent the garden are, on the other hand, more assertive and will unhesitatingly drive away any birds who venture into their feeding area, and I have often seen them prevent bulbuls and mynas and even babblers from flying onto the table feeder if they happen to be already there. The spotted doves, whom Douglas Dewar finds to be 'the most pugnacious and the most pushing' are rather aggressive too, and will brook no nonsense in the feeding area.[5]

The drongos are back! They are especially active during the early mornings and at dusk when their distinctive call can be heard from the trees outside our house; their skilful aerial sallies give me a reason to linger over my tea both at the beginning and close of the day.

21st

Sharad Purnima (*sharad* means autumn) and the full moon is gorgeous. It follows me home, shimmering over the treetops, playing hide-and-seek with the buildings; always brilliant, holding its own in a bright, artificially lit urban landscape. I wake up at four in the morning and it is there, right outside my window, shining through the leaves of the peepal.

Sharad Purnima is celebrated on the full moon of the Hindu lunar month of Ashwin. It is a harvest festival, signifying the end of the rainy season, and is celebrated in various parts of the country. I recall my parents preparing kheer and leaving it out, uncovered the whole night, in the belief that the rays of the moon, on this special night, have healing powers with which they will imbue the rice and milk.

In the West, this is the hunter's moon, so named because this was the time of year when the Native Americans hunted animals to store food for the coming winter. The hunter's moon usually appears in October, except once every four years when it rises in November. Since the lunar months are not in complete synchronization with the astronomical seasons, every three years or so, the hunter's moon shifts to November, and the harvest moon, which falls in September otherwise, moves forward to October. The hunter's moon rises closer to the time of sunset than the moon during the other months of the year, making the light linger longer and the moon itself appear bigger for several evenings in a row. This has to do with the ecliptic—which is the moon's path across the sky—making a narrower angle with the horizon. Further, since the hunter's moon rises near the horizon, making us look through a thicker layer of the atmosphere (which scatters the blue light), it appears bright orange in colour. Magic!

22nd

Phool Walon Ki Sair is a beautiful festival unique to Delhi. The week-long celebrations are held every year in October and are a shining example of communal harmony. It was originally held in the month of August when the monsoon was at its peak; all old written accounts of the festivities add a light 'poohar' or drizzle as a mandatory backdrop to the proceedings.

The festival has a very interesting origin; Akbar Shah's (the second) younger son, who was also his favourite, had a skirmish with the then-British resident over the issue of succession and was banished to Allahabad in 1812. His distraught mother vowed that if he returned, she would offer a chaadar (sheet) and pankhaa (fan) made of flowers at the shrine of Hazrat Qutbuddin Bakhtiar Kaki and a canopy of flowers at the Yogmaya Mandir, both located at Mehrauli. Her wish was granted and the return of her son saw the start of the festival which, except for a gap of a few years, has continued uninterrupted since.[6]

Shehnai players and dancers, along with florists from Delhi, lead a procession to Mehrauli where floral pankhas or fans are offered at both shrines. Many cultural programmes are organized at the Jahaz Mahal, including dances, qawwalis, acrobatics and classical music recitals.

23rd

All this month, the heady fragrance of the harshingar or night jasmine tree perfumes both the evenings and my thoughts. In the morning, the unmistakable tiny white flowers with their bright orange stalks lie strewn on the ground, and if I gather them up to put inside on the table, the scent lingers a little longer. Called 'parijaat' in Sanskrit, harshingar is otherwise an unremarkable tree, small and not meriting a second glance. But come the flowering season and it is spectacular; a real feast for the senses.

The parijaat has a long association with Hindu mythology and is supposed to have come out of the ocean during the great churning or Samudra Manthan (parijaat means that which has descended from the heavens). There are many stories and legends associated with the tree, all of which add to the mystery surrounding it. Buddhist monks use the flower to dye their robes saffron, while in Hindu tradition this is the one flower that can be picked from the ground and offered to the gods.

The flowers have medicinal properties as well and are used to treat certain fevers, inflammation and arthritis. In some cultures they are eaten as part of the afternoon meal.

The fragrance emanating from the flowers is so sweet, so intense that it lingers in the memory long after you have passed under a tree at sunset, and has often tempted me to retrace my steps.

> Set full of leaves [...] among which come forth most odoriferous and sweet-smelling flowers, whole stalks the colour of saffron, which flourish and show themselves only in the night time, and in the daytime look withered and with a mourning cheer.[7]

The pleasure derived from the flowers is no less in the morning when the spent blooms lie scattered on the ground, like stars that fell from the sky during the night, and I find myself almost wanting to believe the legend that says that the parijaat is married to the sun, but he visits her only at night, which is when the flowers bloom forth in happiness, only to fall in sorrow when he leaves in the morning.

24th

Of the two saptaparnis hugging our garden fence, one is smothered in flowers and I am tempted to use the clichéd comparison of being 'bedecked like a bride'. Our evenings are transformed; a heady fragrance fills the house, wafting in even through closed doors, and lingering till daybreak. The tiny florets rain down gently the entire time, transforming the ground underneath into a carpet of mellow cream. They have completely overlaid the seedlings of double-flowered stock and dianthus planted in a narrow bed along the fence. As I sit in my favourite spot on the veranda sipping my morning cup of tea, made all the more enjoyable by the perfume of the saptaparni, the flowers of the harshingar lie scattered in the driveway and the sweet fragrance of the kamini jasmine drifts across from the bush growing alongside. And so the heavily perfumed early morning, with its pastel-hued sky, seduces me into declaring October to be the most fragrant month of the year in Delhi!

26th

Pigeons are the bullies of the avian world, more so even than the crows and treepies. A single pigeon on the feeding table will chase off all birds, irrespective of size, and today morning, I saw one drive away the doves and babblers and mynas. The only ones who are unfazed are the parakeets, who put the intimidators in their place. The bulbuls, especially the red-whiskered ones, are particularly timid and will venture to eat only when there is no other bird on the feeders, serenading us with their beautiful song while they wait. The red-vented bulbuls also break into their joyous chirruping as soon as I put food out, and will try to outsmart the bigger birds who display their hegemony through minatory advances.

I love the antics of the rose-ringed parakeets; after landing on a branch near the feeders, the bird will push himself closer step by step, using his powerful hooked beak to manoeuvre from one branch to the next. The females are bolder and will fly to the feeder directly. The parrots are messy eaters and scatter more grain and seeds than they probably consume! But one bird's riff-raff is another bird's food, and laughing doves and silverbills find the strewn millets and bajra on the ground easy picking.

29th

A beautiful day, blue sky and cool breeze, reminding me of Diana Gabaldon's description of a day in autumn: 'It was a beautiful bright autumn day, with air like cider and the sky so blue you could drown in it.'[8]

Saw my first wagtail of the season—a yellow wagtail—a common winter migrant to the plains from Ladakh and Kashmir. It was running around in typical zigzag fashion in a plot under construction, probably looking for insects disturbed by the digging underway.

> Wagtails remind me of a post-impressionist painting, famous in its day, of a dachshund out for a walk. This picture conveys the gaiety of the occasion in a remarkable way. [...] There are numbers of swishing tails and twirling lengths of lead, all combining to give one the unmistakable impression of eager joy.
>
> Wagtails move their tails in the same token of *joie de vivre*. [...] They move their tails up and down, and a little from one side to the other, in a quick shake, and you see dozens of blurred tails wagging.[9]

Colloquially called 'khanjan', wagtails have been given different names in different areas. As an example, the yellow wagtail, which prefers being near water, is called pani-ka-pilkya, while the white wagtails, which are often seen near dhobi ghats, have been given the amusing name of dhoban. 'Khanjan-eyed' has been used since the olden days to describe someone with beautiful eyes. These birds find reference in several old texts and have traditionally been considered to be a good omen.

November

> 'Early Winter: Abhinanda
> The round villages are charming now at day's end
> with threshing circles scattered on the common
> for treading of the heaped up rice;
> the dung fires cast a ring of smoke
> that hangs low overhead from weight of frost.'
>
> —VIDYAKARA, *Subhashitaratnakosha*

1st

November is a beautiful month in Delhi, a sort of transition between autumn and winter. The temperature hovers between an average high of 28°C and an average low of between 15°C and 18°C. It is a sunny month with hardly any rainfall unless western disturbances bring unsettled weather. Diwali usually falls in this month, making it a time for celebration and festivities. The average wind speed in Delhi is the lowest in winter, and since the city is anyway not very windy, the days can sometimes feel totally still. The months of October and November, when the wind speed is the least, are therefore often smoke-filled and hazy with high pollution levels. Add to this Delhi's location, north-east of the Thar Desert and south-west of the Himalayas, forming a sort of bowl where pollutants get trapped. Dust from the arid zone surrounding the city has no outlet for escape and blights the entire Indo-Gangetic plain from Punjab to West Bengal. What further aggravates the situation is the fact that this month has an almost negligible amount of rainfall. The length of day is steadily decreasing as the month progresses, leading to a nearly 35-minute difference between the 1st and the 31st. The level of humidity also remains low throughout as does the cloud cover.

Looking at the heavens, November brings three planets to the evening sky—Venus, Jupiter and Saturn—which towards the end of the month will stretch across the south-western sky in a 35-degree-long arc of the ecliptic.

2nd

A very pleasant start to the new month; a lesser whitethroat on the bamboo fence that marks the boundary of our garden, and a small flock of chiffchaffs high up in a nearby gulmohur. The whitethroat spent a good 10 minutes at the same spot and was still there while I hurriedly fetched the binoculars. This small grey-and-white passerine may be rather unremarkable to look at but has just completed a most remarkable feat; it has flown all the way from temperate Europe to winter with us, thus making a round-trip of several thousand kilometres. Weighing a mere 12 gm, the birds start preparing for their long journey in autumn by augmenting their chiefly insectivorous diet with berries.

Bird migration is one of the most enduring of nature's miracles; nearly 4,000 or 40 per cent of the world's 10,000 or so species migrate, flying over thousands of kilometres, often at great altitudes. According to the National Audubon Society, bar-headed geese reach altitudes of up to five and a half miles above sea level while flying over the Himalayas, and even more astonishingly, Arctic terns can fly nearly 50,000 miles in a year, clocking up three trips to the moon and back in their lifespan of 30 years![1] One hundred and sixty-five or nearly 12 per cent of the 1,395 bird species of the Indian subcontinent are winter migrants.[2] As mentioned earlier, India lies on the main Central Asian Flyway, one of the nine flight paths used by migrating birds worldwide. Interestingly it also lies at the junction of two other major flyways of the world; the East Asian–Australasian and the Asian–East African.

> Some birds like Pied Crested Cuckoo […] and Amur Falcon […] besides travelling through the Central Asian Flyway, also pass through the West Asian–East African Flyway. Similarly, birds travelling along eastern India, Bangladesh, and Andaman and Nicobar Islands, such as Red necked Stint, […] Spoon-billed Sandpiper […] and Spotted Greenshank also pass through the East Asian–Australasian Flyway.[3]

Bird migration is, however, just one part of a magical global exchange that is continuously taking place all across our planet and at all times, often without us being aware of it. Winds move across countries and continents. Jet streams blow strongly, like a river of air, in the upper reaches of the atmosphere, circumnavigating the entire Earth. The waters of the ocean flow horizontally as currents and vertically as upwellings and downwellings, creating a huge intercontinental exchange of heat and nutrients. Further, waters of different densities intermingle, reducing differences and creating the global conveyor belt. Tidal migration, driven by a primal desire for food and mating, compels organisms to often travel hundreds of kilometres; salmon migrate between saltwater and freshwater every year, while tortoises and several species of fish traverse long distances. Butterflies and dragonflies fly across continents, reindeer journey more than 4,000 km per year in North America, the Serengeti migration sees millions of animals move over long distances annually and the wizardry of nature continues!

5th

At a 15-minute walk from where we live is an enclosed piece of land, government-owned, with restricted entry. Situated at a slight elevation on a rocky outcrop, it is surrounded by a paved, shady walking path. A pleading look will sometimes get you in, depending, of course, on the chowkidar's mood; even otherwise, it is a rich hunting ground for twitchers like me; the chicken coop fencing allowing for uninterrupted views of the goings-on inside.

Because of the underlying rocky surface, the vegetation is semi-arid, mainly shrubs and grasses, interspersed here and there with trees. On this morning in early November, the grasslands were alive, swaying to the music of the breeze, golden fronds nudging each other softly, the perfect playing field for a pair of common stonechats, pied bush chats, paddyfield pipits and the ever-present bulbuls. The glasses are tall at this time of the year, a perfect camouflage for the crested lark, which I saw only when it rose high in the sky. I confess to being confused about its identity, this being my first sighting of this bird, and it was only when I returned home and consulted the bird book that I was able to finally rejoice!

Another wonderful sighting was that of a family of scaly-breasted munias, parents and juveniles, on a clump of dried grass very near where we were walking. They were pecking on some seeds that lay scattered, and while we stood—within five minutes—they were joined by friends and relatives, including a family of black-headed munias. A pair wrestled with what looked like a small ball of tangled fibres, the male lifting it with his beak and hitting it on the road, presumably in a bid to separate some strands. The scaly-breasted munia is an occasional visitor to our garden too, but always as part of a flock of silverbills and house sparrows. It is a small bird with a list of names longer than itself; nutmeg mannikin, spice finch, tilyar or spotted munia and checkered munia!

6th

The morning resonated with the loud screeching of a small flock of plum-headed parakeets, who landed on the topmost branches of a nearby tree in a perfectly synchronized manner. The morning sun turned their green into gold and accentuated the red of their beaks and the plum of their heads, making them shine in multi-hued brilliance.

7th

On a windy, sunny winter's day like today, serenaded by birdsong and surrounded by flowers, there is nowhere else I'd rather be. My little corner with its dappled light and shade seems so far removed from the hustle and bustle of a big city.

We have a tiny new visitor to the garden, and he is adorable; an ashy prinia who, for the past week or so, has dropped by every day around noon. He is a bold little chap and doesn't hesitate to forage in the pots of chrysanthemums near where I am sitting, working his way through all of them one by one before shifting his attention to the small insects and bugs that might be found in the adjoining flower beds. I notice that even though his movements are swift as he flits about in a very busy manner, flicking his tail all the while, he does like to take his time and linger for a bit in every patch, repeating his 'zeet zeet zeet' song as he goes about his work.

This small ashy, white and buff-coloured passerine is endemic to the Indian subcontinent, and equally at home in the urban landscape as in the countryside. An interesting feature of prinias is the snapping sound they make during flight, thought to be produced by either the wings or the tail.

> How this noise is made we do not know. It sounds as though it were due to the upper and lower mandibles of the beak closing sharply together.[4]

10th

A most interesting early morning sighting. Whilst driving along the East Delhi highway, where lots of construction work is in different stages of completion, we came across a small stretch that had buildings with only two floors finished. The roofs, in all cases, were laced with the vertical iron bars that would form the next elevation, but which, for now, housed a colony of black kites. Each of the bars, and there must have been more than 50 per roof, had a kite perched on it!

Later in the evening, an absolutely beautiful sunset; not so much fiery oranges and reds as soft peaches and pinks. A band of grey clouds hung low over the western horizon, sandwiched between Earth and sky, making for a perfect contrast with both. The gathering dusk, illuminated by a three-day-old crescent moon with a bright Jupiter right next to it, formed a tableau such as an artist might have painted.

In the waxing crescent phase, the moon is just out of the new moon stage and is dark, except for its right edge, which appears crescent-shaped. As the days pass, the moon will reach the first quarter, when half its surface will be illuminated, before gradually moving on to becoming a waxing gibbous with more than half its surface sunlit. And then finally we have the glorious full moon!

Fun fact; if you're out hiking and want to know whether the moonlight will increase in the coming days, thus illuminating your path further, and if by chance you happen to be in an area where your phone has no signal, worry not! There is a very easy way to tell whether the moon is in its waxing (growing) or waning (reducing) period. If the moon is visible at sunset it is waxing; the waning moon rises later at night. Also in the northern hemisphere, the illuminated part or horns will be to the right and the shadow part to the left.

I sometimes think about my interest in all things celestial, but then I tell myself that I am not alone in this wonderment. '…we need not feel ashamed of flirting with the zodiac. The zodiac is well worth flirting with.'[5]

12th

My husband and daughter, back from a late afternoon stroll in the shady lane behind our house, reported seeing an unfamiliar bird, which they had managed to photograph. It was an orange-headed thrush, which was still there skulking in the undergrowth when I reached five minutes later. However, it almost immediately flew to an adjoining tree, settling on one of the higher branches, where it stayed motionless for several minutes.

The orange-headed thrush is a winter visitor to Delhi, migrating down from the Himalayas. Such a striking bird; orange and blue never looked so good together!

14th

The morning sky is a bright cerulean blue across which streaks of white clouds fly, pushed along by a strong wind, which creates a brouhaha among the leaves of the peepal, causing them to rain down in a golden shower. Later in the afternoon, the sky becomes a choppy sea, with white foam-crested waves falling over themselves. Disappointingly, however, there is no rain.

Returned home after two days from an outstation trip to find that the peacocks had eaten up all the flower seedlings planted nearly a fortnight ago; the dahlias and double-flowered stocks had their tips chopped off, while the rows of candytuft and alyssums were totally empty, baby plants probably pulled out and snacked on.

I was rather annoyed with the blighters because I will have to go and buy the seedlings again, and very often many are sold out by this time of the month. But the beautiful big birds are still my friends, and I don't want to banish them from the garden. It is also clear that keeping seeds out for them is not sufficient to deter them from wanting to nibble on some fresh greens. A physical impediment was the need of the hour and, after much deliberation, I decided to get medium-height bamboo enclosures made around the flower beds. These are eco-friendly and reusable; the bamboo sticks are laced together with a jute rope and can be folded and put away once the plants are bigger.

17th

A double bonanza: an early morning sighting of a pair of Marshall's ioras, and then a few minutes later, a pair of plum-headed parakeets, both outside the temple in our residential complex. The ioras were frolicking high in a gulmohur tree and it was only the reduced density of leaves at this time of the year that enabled me to enjoy their antics. However, they showed an almost lamentable disregard for my convenience, flitting from branch to branch at such speed to make me dizzy from just trying to keep up with them. I wonder how they manage to eat at all—they can't possibly be gobbling up insects and flying about at that rate simultaneously!

These small black and yellow arboreal birds are easily confused with the common iora and it was only when one descended to some bushes growing at the side of the road that I was able to spot the confirmatory white edging to the black tail. All the time the birds were in the vicinity, flitting from branch to branch, often hanging upside down to reach an elusive morsel, the pair kept in touch by a delightfully musical whistling that was reminiscent of bells ringing faintly in the distance.

In Madhya Pradesh, where we lived for many years, I had the occasion to observe the common iora at many locations; in groves of neem and peepal on the outskirts of villages; in the huge gardens of the houses we lived in; in the secluded exclusivity of circuit houses with their ancient trees. I confess to being somewhat confused at times, for the birds sounded like ioras with their onomatopoeic rendering of 'shaubeegi' (their Hindi name) and behaved like ioras, but had slight variations in colour and markings. Some research revealed that they were probably different races, and since the places we lived in were often at the cusp of either Central and South India or North and Central India, there was a possibility of some overlap in the birds' ranges.

I remember how amused I was to observe, for the first time, the theatrical display of the male during courtship. It was towards the end of July; the monsoon had broken over Central India a fortnight earlier, and the early afternoon was overcast. As I ventured outside to assess the condition of the waterlogged flower beds, I saw, on a low-hanging branch of a sal tree that grew in a corner of the garden, a pair of common ioras. The male, in full dress regalia, would fly up from where the female sat and fluffing up his back feathers would perform an astonishing spiral drop in slow motion as if he were descending in a parachute. He did this repeatedly, serenading her all the time with a beautifully whistled tune, but unfortunately she was unimpressed, and flew off to another tree soon after, hotly pursued by our disappointed, but not discouraged, friend!

19th

Full moon night! The November full moon is called Kartika Purnima in the Hindu calendar and is considered to be the most auspicious day by the Hindus, Jains and Sikhs. This full moon is also called Tripurari Purnima and Deva Diwali by the Hindus and is marked by prayers and religious worship.

Jains worship this day by visiting Palitana, a pilgrimage centre located in Gujarat. This walk up the Shatrunjaya hills, which has more than 3,000 temples, is a very important event in the life of a devotee. For the Sikhs too, this is a very important day, being the birthday of Guru Nanak Dev, and is celebrated as Gurupurab.

In the Western calendar the November full moon is called the beaver moon, this being the time of year when beavers prepare for the cold season ahead by building their dams called 'lodges'. Since beavers are nocturnal animals, they work in the moonlight, thus giving their name to the full moon. The November full moon is also called the frost moon and if it is the last full moon before the winter solstice, it is also called the mourning moon.

20th

A beautiful late autumn day, overcast with the sun putting in an occasional gander from between low-slung stratocumulus. The first of the winter flowers are just beginning to bloom; petunias and dianthus and tiny alyssums. But it is the chrysanthemums that are the stars of the show this month, and the pots that I planted some months ago fill out some more each day, bringing colour to an otherwise in-waiting garden. It is so exciting when the first tightly closed buds open up a bit and one can make an informed guess about the colour.

Garden centres and nurseries are filled with chrysanthemums of different types and colours this month, and one is truly spoiled for choice. My favourites are the pom-pom or button mums and the aptly named spider chrysanthemum, whose thin, long petals look exactly like spider legs. Our two pots are a beautiful shade of yellow-orange and maroon. We also have pretty daisy mums, which look very much like

daisies, only with multi-layered petals. I have been looking for chrysanthemums with long pointy petals that look like quills but with no success. I did, however, get two unusual-looking mums, with the petals of the flowers drooping away from the centre. They come in gorgeous colour ways and I chose one with red petals with a yellow stripe running down the centre and another that has white petals with a purple stripe. My mother liked to make chrysanthemum tea for her mid-morning beverage, and for this, she would pluck a fresh bloom every morning from her rather considerable collection. I never did develop a taste for the brew, for which I think she was secretly relieved, probably not wanting to pluck a second flower!

24th

Today afternoon at around half past three, when the sun had retired from our little corner for the day and the wind had turned chilly, an astonishing thing happened. I had spent a lazy hour watching the doves and silverbills forage under the feeders where the parrots, who are very messy eaters, had dropped considerable quantities of bajra and jowar, and was contemplating shifting indoors and having a cup of tea when a shikra swooped suddenly, as fast as lightning, and attempted to grab a laughing dove who was feeding some distance away from the others. The raptor passed just above my head, startling me into getting up jerkily, which, in turn, startled the dove, probably saving its life. In an instant the garden was empty and lifeless; the shikra settled on an overhanging branch of the saptaparni, plotting his next move, while the babblers, who are the guards of the avian world, kept up a huge cacophony the entire time.

The harshingar has finished flowering, but the promise of fragrance continues with the saptaparni, which is still in bloom in some localized patches, and the *raat ki rani*, whose perfume hangs low on the cold air. The berries on the neem and the Persian lilac are a nutty brown in colour and look pretty against the blue of the cloudless autumn sky. Of the two saptaparnis growing right next to each other at the front of our house, the one that flowered profusely has thinner, yellower, less robust leaves than the other which did not bloom much, as if the blooms sucked out the very essence of the plant. Visited Chandigarh over the weekend and saw a very interesting tree—the flamegold rain tree—which I have somehow missed spotting in Delhi. Endemic to Taiwan, this is a common tree in tropical and subtropical regions. It flowers in summer, after which the three-lobed fruit forms and stays on the tree for long, and makes for a very pretty sight. The tree I saw was smothered in pretty rose-coloured inflated papery cases that looked like miniature Chinese lanterns bunched together like grapes.

29th

Ended the month with the wonderful sighting of a Hume's warbler in the overgrown, shady lane behind our house. This tiny bird, named after Allan Octavian Hume, a famous ornithologist and British civil servant, is a winter visitor to the city from its breeding grounds in the Himalayas, and is rather difficult to spot, with its greyish-olive colouring blending perfectly in its preferred habitat of tree canopies. In this case, it was the bird's distinctive buzzing call 'chiz-it chiz-it' that alerted me to its presence in a dense thicket of shrubs, where it was hunting with a small mixed flock of birds. Hiding behind a tree, I stood absolutely still for several minutes, watching it flit from branch to branch; restless, very active, flicking its tail from time to time. W. Eugene Oates, writing in 1889, says of the bird[6]:

> A winter visitor to the plains of India [...] found throughout the Himalayas as far as Nepal, but it has not occurred in Sikkim. [...] It is known to breed abundantly in Kashmir, and probably its migration does not extend beyond the Himalayas.

And so,

> '*November comes*
> *And November goes, […]*
> *With night coming early,*
> *And dawn coming late.*'

—ELIZABETH COATSWORTH, Twelve Months Make a Year

30th

Poinsettias come into the market at the very beginning of the winter season, and by the end of this month, they are available in most garden centres. I love them and get some plants every year. This time I have one red, one orange, two pinks—a dark one and the other blush-toned—a gorgeous cream-coloured one, a variegated red and white and one with curly-twirly petals.

December

'Comes the bright and bracing winter to the royal Rama dear,
Like a bride the beauteous season doth in richest robes appear,
Frosty air and freshening zephyrs wake to life each mart and plain,
And the corn in dewdrop sparkling makes a sea of waving, green, [...]
Southward rolls the solar chariot, Himalaya, "home of snow",
True to name and appellation doth in whiter garments glow,
Southward rolls the solar chariot, cold and crisp the frosty air,
And the wood of flower dismantled doth in russet robes appear!'

—ROMESH DUTT, *The Epic of Rama, Prince of India*

2nd

December is a beautiful month in Delhi, a time for cold clear days, the January fog not having set in yet. The temperature hovers between a low of about 9°C and a high of 20°C or so, making for an average of about 12°C. Afternoons are sunny and bright with rainfall limited to a couple of days at the most. Technically, days start getting shorter and the nights longer after the summer solstice in June, but in a city like Delhi situated in the subtropical zone, the change is barely perceptible till the end of October. That's when I suddenly noticed that the dusk is coming in much earlier and it is dark by 6 o'clock, the trend continuing till December, which is the month with the shortest daylight period of 10 hours.

The festive mood that steeps into the city before Diwali flows through December as well, continuing right up to the start of the new year. Markets are strung with fairy lights, shops decked out in torans, Santa toys belting out 'jingle bells', 'Happy New Year' banners available at every nook, baskets filled with chips and chocolates and other treats, heaped outside grocery stores. Add to this the buckets of fresh flowers at every florist's and the delicious mince pies and plum cake on sale at every good bakery!

6th

A cold, overcast winter day, but a light wind has picked up in the last half hour, driving away the last of the morning fog. The leaves of the goolar, dry and brittle, clatter down noisily with the slightest breeze.

Since morning I have been noticing a battle for supremacy between two warring clans of jungle babblers and as I write this, there are more than 20 of them in the garden, keeping up a constant noisy conversation punctuated with much squeaking and scolding. A fragile peace has been brokered for the time being, after a long bitter fight between rival gangs, complete with tumbling, screeching, falling to the ground, mid-air sorties, et al.

Just after sunset, towards the west, I observed a beautiful crescent moon (waxing) very near a spectacularly bright Venus. December this year will see the moon conjunct with five planets, and while it will not be possible to observe all, I hope to see the moon–Saturn conjunction after two days and the moon–Jupiter on the day after that. Last year on the day of the winter solstice, the planets Jupiter and Saturn were the closest to each other they have been in the last 800 years, just a 10th of a degree apart when viewed from Earth. The light from these planets combined to form a bright point, creating the same star of Bethlehem that guided the three wise Magi to the birthplace of the infant Jesus!

8th

Lots of butterflies these days! They start visiting just before noon when the garden is drenched in sunshine, and the flowers have turned their faces fully to the sun. Today, seeing an unfamiliar butterfly flitting about with a tortoiseshell and a couple of common gulls, the amateur lepidopterist in me rushed inside to consult my copy of *The Book of Indian Butterflies* that informed me that the stranger was a Danaid eggfly, common in the region.[1] To my surprise, I also learnt that the tortoiseshell-lookalike was actually a female eggfly!

December is the month for poinsettias and I always get them in different colours. The only problem is that they don't repeat flower the next year even if you manage to keep the plant healthy through the summer. I did that, and although it grew bushy come winter, the leaves did not change colour even in our coldest spells.

9th

After the near doldrums of October and November, the wind has picked up slightly, leading to clear blue skies. It is not much movement of air really, but it feels like a lot and my rustic wind vane is going crazy. It's a dried leaf that got caught in a piece of twine some weeks back, and now hangs in a corner of the bamboo fence and twirls crazily in even the slightest breeze. It hardly moved at all the entire last month, a final proof of the suffocating windless days we've had.

For the last couple of days, we have had new guests at the feeders; a pair of white-cheeked bulbuls, the least common of the bulbuls in our area. They do visit the garden sometimes, though not very often, and this is their first time at the bottle feeders. They are shy birds, flying up at the slightest disturbance and settling on the overhanging branches of the allamanda. I have tried to lure them by scattering some bajra on the ground in a secluded corner near the water pot. Let's see if it works.

12th

Went out for lunch. One of the pleasures of winter in Delhi is the opportunity to eat outdoors, under a blue sky, in one of the parks or eateries. The other delightful things about the season are the nargis flowers, which are available only for two months, and plum cakes. The nargis is my favourite among flowers, with its fragrance redolent of lazy winter afternoons in my mother's garden where they bloomed interspersed with other winter flowers. Other bulbs that do well in this season are gladiolus and freesias, and I have a couple of pots of each. The freesias are doing rather well this year and shine jewel-like in their big terracotta pots.

At my favourite café, sunlight slants in from under the half-open window next to where we are sitting, diffusing through the bottles of coloured glass on the windowsill and settling like a golden blanket on the nargis I just bought. There is something special about winter sunlight, a mellowness that alchemizes everything the light touches, so that flowers glow jewel-like, the blue flash of a kingfisher is more vivid and the drying leaves of the neem appear gilt-edged. On the way back home, we saw a pair of Indian pond herons at the fountains of Vijay Chowk, standing so still as to be motionless. This common Delhi resident can be seen around most waterbodies and has adapted well to the urban landscape it calls home. Also at the same Chowk, running around in the adjoining gardens, was a grey wagtail, oblivious to the constant motion of a city going about its business.

13th

Very cold and foggy; the day before, the midday temperature was only about 2°C lower than the night minimum, making it a full 10°C lower than normal for this time of year. A light drizzle towards the evening made the fog even denser, wrapping the dusk in a silent white coat. The drongos, who seem to love such days, have kept up their lively antics throughout, making aerial sallies from their perches in the nearby trees, briskly going about their business of chasing after insects.

In the evening, the sunset is spectacular, not in a showy, flamboyant way, but gently in pastel-hued beauty. The sky is salmon pink, arranged in distinct bands, darkest at the horizon, fading to an almost baby pink overhead, against which the silhouette of the trees stands out darkly, their bare branches systematized like a dendritic river system.

16th

A beautiful winter's day, windy and sunny, with a cloudless

cerulean blue sky. The flowers in the garden look happy, the birds are tweeting about their business, and the lower branches of the goolar, which were bare up to a week ago, are clothed in new leaves that are refulgent as only fresh growth can be. I find this surprising because traditionally new leaves appear on the goolar only in March, but then I notice the same phenomenon in other goolars in the neighbourhood, so maybe it has something to do with climate change.

The sun is casting dappled shadows on the walls and the wind is making them dance to its tune. Of all the trees, the shadow of the peepal is the prettiest, a faithful impression of the perfectly heart-shaped leaves, while that of the neem is more of an indistinct blur. The upper branches of the goolar are still bare and form criss-crosses on the cobbles below.

17th

Windy, sunny days like today have a timeless feel to them; the winter sun may be mellow but it limns each corner and gilds each surface with gold. Middays are warm so that it is still summer in the sun and winter under the shade of the peepal and neem.

A bird which doesn't seem to mind either the chill of daybreak or the cold of late evening is the black drongo. In the fog of the early morning, while the visibility is still very low, it is already about its business, making aerial sallies from its perch, appearing almost ghost-like in the gloom, a dark shadow spelling doom for its victims. In the evening too, black drongos hunt till late, often well into the gathering dusk, and many times I have observed the more intrepid among them making a feast of the insects that swarm to the street lights at night. They are noisy on the wing, and become very vocal while returning to their perch, uttering a harsh 'tchee-tchee'.

These are skilled hunters, sitting motionless on semi-exposed branches of trees until they sight their prey, when they strike swiftly and purposefully. These slim birds are aggressive and pugnacious and totally fearless, and E.H.A. says of them, in his characteristic humorous way[2]:

> The king-crow [...] leaves the whole bird tribe far behind in originality and force of character. [...] He does not come into the house, the telegraph wire suits him better. Perched on it, he can see what is going on, and keep all the other inhabitants of the compound in order. He drops, beak foremost, on the back of the kite, levies the tribute of a feather from the passing crow, and jeers the blue jay as it goes rolling by, like a ship in a heavy swell, with a lazy flapping of its rainbow-coloured wings. Anon he spies a bee-eater capturing a goodly moth, and, after a hot chase, forces it to deliver up its booty.

Black drongos are excellent mimics and have a huge repertoire of calls, which they use either to steal food or scare away birds bigger than themselves. They often mimic a shikra to scare smaller birds into flying away, leaving their food behind! No wonder their Hindi name of kotwal!

Black drongos are year-round residents of Delhi, but I have seldom spotted them during the hot summer months. Maybe some of them migrate to the south to breed, because, despite my deliberate vigil, I don't recall coming across any drongo nests in the vicinity. However, come winter and suddenly their distinctive call is once again part of the landscape and a single bird can then make up an entire orchestra. A drongo that frequents the saptaparni in front of our house can keep up its performance for a full five minutes at a time; a harsh 'chhee' followed by a short melodious whistle, the bird whistling and rasping alternately.

18th

A befittingly brilliant last full moon of the year, orangish-red and otherworldly. It rises low and large in the late evening, barely skipping the tops of the trees, but when I wake up a little after midnight I find it riding high in the sky and am grateful to be touched by the magic light that makes the urban landscape a little less dreary.

The moon nearest to the winter solstice is called the full cold moon because it signals the start of the coldest time of the year. It is also known as the frost moon or the winter moon and the long night moon because it rises near the longest night of the year and also because this moon stays above the horizon for a longer time than during the rest of the year. According to the Hindu calendar, this full moon is called Margashirsha and is considered to be especially auspicious. Devotees keep a fast till moonrise and bathe in one of the holy rivers.

21st

The winter solstice.

> 'This is the solstice, the still point
> of the sun, its cusp and midnight,
> the year's threshold
> and unlocking, where the past
> lets go of and becomes the future;
> the place of caught breath, the door
> of a vanished house left ajar...'
>
> —MARGARET ATWOOD, *Eating Fire*

I don't know whether I love or dislike this day more. Love it because it is midwinter for us here in subtropical Delhi, and cold crisp days when the dark comes in a little earlier are my absolute favourite. Dislike it because the shortest day and longest night of the year signal a shift in the movement of the sun, which has now reached the most southerly point of its journey. As it retraces its steps, the days will start getting longer and the nights shorter for us in the northern hemisphere, and slowly but inexorably, summer, which has been waiting in the flanks, will be here again. The thought of the endless hot summer makes me anxious, and I find in myself a deep desire to hold on to the cold a little bit longer. However, I did follow my childhood ritual of checking for my noontime shadow, which is the longest today in the entire year. But disappointingly, due to the sky being slightly overcast, and the trees and the buildings interfering with the sunlight, I could not get a clear impression.

In Delhi, the December solstice is 3 hours and 39 minutes shorter than the June solstice. However, due to a concept called the 'equation of time', which is basically the discrepancy between the time measured by the sun's daily movement across the sky called the 'apparent solar time', and the steady time that modern day clocks measure, called the 'mean solar time', the earliest sunrise and latest sunset do not occur on 21 December. The former happens a few days later and the latter a few days earlier than the solstice.

> On Christmas Eve and Christmas, the sun rose a minute later than the solstice but also set a minute later. It continued to rise and set at the same time till the 27th. It appeared that a minute added in the evening was evened out by a minute's tardiness in rising. On the last day of the year, the sun set at 5.29 p.m.—the same as it did on the solstice.[3]

24th

The shady lane behind our house is a wonderful place to see birds; cobbled, lined with trees and almost always deserted, with overgrown shrubbery and undergrowth. Today being bright and sunny, I decided to take a casual stroll in the dappled sunlight, adding my shadow to the ever-shifting patterns formed by the leaves overhead, and on the very first tree outside our gate, I spotted a grey-headed canary flycatcher flitting about the canopy along with some chiffchaffs. Behaving in a true flycatcher manner, jerking its tail and performing incredible aerial sallies, the bird entranced me for a good 10 minutes. The grey-headed canary flycatcher is a winter visitor to Delhi, and my previous sightings of this bird have all been in the hills.

Chiffchaffs are also winter visitors from the Himalayas and are the quintessential small, plain brown birds, being no bigger than a sparrow, with dull whitish underparts and front. But what they lack in colour, they make up in sprightliness, zipping through the canopy, hopping on the ground, foraging in small shrubs, wings flicking and tails jerking in typical warbler fashion!

26th

Yesterday being Christmas and a holiday, we decided to go to Chandu Budhera for some birdwatching. Located just 10 km from Gurgaon and very near the Sultanpur Bird Sanctuary, Chandu is a small hamlet, part of the larger Najafgarh wetland, which straddles the Delhi–Gurgaon border. These marshes are home to a huge variety of both resident and migratory birds, the Najafgarh Jheel being the second largest waterbody or wetland in the Delhi region.

Because of the cold and the fog, which often lasts until noon, we left for Chandu just after lunch. It was a bright sunny afternoon, with a golden light accentuating every colour and feature of the landscape. Earth and sky teaming with birds; waterbirds and birds of the marshes and fallow fields, and overhead the sky swarming with red-rumped swallows, barn swallows, wire-tailed swallows, plain martins and little swifts. All darting about, creating invisible criss-crossing patterns in the air.

Our birding was restricted to a small area that consisted of farmland and fallow waterlogged fields bounded on the margins by tall grasses and reeds. The marshlands were overrun by migratory waterbirds, while the reeds were home to a number of warblers and prinias. As we entered, we were greeted by the loud calls of the clamorous reed warblers and the softer notes of the moustached warblers, although both proved rather difficult to observe and we had to wait in the vicinity for several minutes before we could spot either. Both are winter migrants to the plains, much like the paddyfield warbler who flew low across the reeds just in front of us, giving us a near-perfect photograph. Another wonderful sighting was that of a Blyth's reed warbler a short distance away.

The resident birds were out in full force too; three types of prinias—ashy, plain and graceful, the last with its gorgeous tawny eyes—pied bush chats, a family of red munias, silverbills, common and white-throated kingfishers and long-tailed shrikes. We spotted a number of herons, including an entire colony of black-crowned night herons. We were also able to photograph grey and purple herons and intermediate egrets as well as Indian pond herons. Purple swamphens shone sapphire in the golden light while the common moorhens kept up a constant chatter. A great bittern flew past nearby, affording us a fabulous view!

Among the waterbirds were common coots, ferruginous ducks, gadwalls and little grebes. Serendipitously, in a single stretch of road along a dry man-made water channel, we saw five types of wagtails, all winter visitors to our city—white, white-browed, citrine, yellow and grey. Wagtails are smartly liveried little birds; walking purposefully in their quest for food, making a run for any insect that tries to get away, calling on the wing when disturbed, all the while wagging their tails up and down. As we were walking back to the car, I spotted a pair of streaked weavers in the tall reeds and grasses growing along the road. This was my first sighting of these birds and I was charmed! But there were more surprises still; a small flock of crested larks and a lone paddyfield pipit in a dry, scrub-covered piece of land. And finally a pair of bluethroats, visitors from a long way off!

31st

As I bask in the mellow winter sunshine on this last day of the year, a chocolatey mocha warming my cold hands, my heart is full and my cup of contentment overflows. A light breeze rustles in the trees, flowers bloom all around me and birdsong fills the air. A beautiful peacock visited us this morning and a short while ago a crow pheasant dropped in for a mid-afternoon snack. From where I am sitting, I can see a flock of Indian silverbills busy on the smaller feeders, while a dozen or so rose-ringed parakeets are trying to pry peanuts and corn from the bigger bottle-shaped ones. Sparrows flit around, feeding on the grains and millets I scatter for them on the ground every day and I am happy to note the considerable increase in their numbers. All three of our resident bulbuls—red-vented, red-whiskered and white-cheeked—frolic alongside, hopping from bush to food and back again, as do the magpie-robins, who have started serenading us occasionally these days. A treepie, surprisingly nervous for his size and reputation, drops by every afternoon looking for titbits, and I like to put out biscuit crumbs for him. A pair of tailorbirds has become rather friendly as has an ashy prinia who visits alone. The babblers are so tame that my husband feels they would be willing to eat from my hand, but I am not going to try that just yet. I am chuffed at the thought that I have a personal friend in a crow who flies over whenever I'm in the garden. The number of laughing doves who frequent our little corner has grown from one to nearly a dozen. They are now joined by a pair of spotted doves as well as Eurasian collared doves. A pair of sunbirds has made our home their own, while a couple of pied mynas have become rather tame in the space of a few months, and often hop over to where I am sitting. They are rather aggressive birds and do not hesitate to scold the common mynas who vastly outnumber them and who can be equally quarrelsome. The brahminy mynas who visit us are, however, more circumspect and like to keep their distance, much like the flock of Oriental white-eyes that comes only to the water pot and is never tempted by the feeders. Other birds who like to pay a call from afar are the black-rumped flameback woodpeckers, grey hornbills and koels. The brown-headed barbet is a regular visitor to the garden, although the coppersmith keeps his distance. Just yesterday, a large green barbet deigned to come down from his perch on the saptaparni when I accidentally spilt a considerable amount of grain on a ledge overhanging a window, but the visit was rather brief.

 The only trouble in paradise comes in the form of two shameless cats, one small black-and-white and the other a bigger tawny, who are forever trying to snare the squirrels and birds. But then, *c'est la vie*, that's life!

List of Trees

African mahogany	*Khaya senegalensis* (Desr.) A.Juss.
African wattle	*Peltophorum africanum* Sond.
Amaltas (Indian laburnum)	*Cassia fistula* L.
Anjeeri (Punjab fig)	*Ficus palmata* Forssk. subsp. *virgata* (Roxb.)
Arjun	*Terminalia arjuna* (Roxb. ex DC.) Wt. & Arn.
Babool (Egyptian thorn)	*Acacia nilotica* (L.) Del. subsp. *indica*
Bakain (Persian lilac)	*Melia azedarach* L.
Banjh (Grey oak)	*Quercus leucotrichophora* A.Camus
Banyan/Badh (Strangler fig)	*Ficus benghalensis* L.
Baranimbu (Lemon)	*Citrus limon* (L.) Burm. f.
Bhendi (Indian tulip)	*Thespesia populnea* (L.) Sol. ex Corr.
Caribbean trumpet tree	*Tabebuia aurea* (Manso) Benth.& Hook.f. ex
Champa (Frangipani/temple tree)	*Plumeria obtusa* L. var. *obtusa*
Chamrod	*Ehretia laevis* Roxb.
Chikrassy (Indian redwood)	*Chukrasia tabularis* A.Juss. var. *tabularis*

Dhak/Palash (Flame of the forest)	*Butea monosperma* (Lam.) Taub.
Dhau	*Anogeissus pendula* Edgew.
Earpod wattle	*Acacia auriculiformis* A.Cunn. ex Benth.
Eucalyptus (Lemon-scented gum)	*Corymbia citriodora* (Hook.) K.D.Hill & L.A.S.Johnson
Flamegold rain tree	*Koelreuteria elegans* (Seem.) A.C.Sm.
Floss-silk tree	*Ceiba speciosa* (A.St.Hil.) Ravenna
Glaucous cassia	*Senna surattensis* (Burm. f.) Irwin & Barneby
Goolar (Cluster fig)	*Ficus racemosa* L.
Gulmohur (Flame tree)	*Delonix regia* (Boger ex Hook.) Raf.
Harshingar (Night-blooming jasmine)	*Nyctanthes arbor-tristis* L.
Hingot (Desert date)	*Balanites roxburghii* Planch.
Hong Kong orchid tree	*Bauhinia* x *blakeana* Dunn.
Imli (Tamarind)	*Tamarindus indica* L.
Indian coral tree (Tiger's claw)	*Erythrina variegata* L.
Jaggery palm (Toddy palm)	*Caryota urens* L.
Jamun (Indian blackberry)	*Syzygium cumini* (L.) Skeels
Jarul (Queen's crêpe myrtle)	*Lagerstroemia speciosa* (L.) Pers.
Jatropha (Purging nut)	*Jatropha curcas* L.
Jhand	*Prosopis cineraria* (L.) Druce
Jhinjheri (Burmese silk orchid)	*Bauhinia racemosa* Lam.
Jungle jalebi (Monkeypod)	*Pithecellobium dulce* (Roxb.) Benth.
Kachnar (Mountain ebony)	*Bauhinia variegata* L. var. *variegata*

Kadamba (Common bur-flower)	*Neolamarckia cadamba* (Roxb.) Bosser
Kaner (Yellow oleander)	*Thevetia peruviana* (Pers.) K.Schum.
Kaniar (Orchid tree)	*Bauhinia purpurea* L.
Kanju (Indian elm)	*Holoptelea integrifolia* Planch.
Kapok (White silk cotton)	*Ceiba pentandra* (L.) Gaertn.
Karanj (Indian beech)	*Pongamia pinnata* (L.) Pierre
Kareel (Bare caper)	*Capparis decidua* (Forssk.) Edgew.
Kassod (Yellow cassia)	*Senna siamea* (Lam.) Irwin & Barneby
Katsagon	*Fernandoa adenophyllum* (Wall. ex G.Don) Steenis
Khair (Black cutch)	*Acacia catechu* (L.f.) Willd.
Kosam (Lac tree)	*Schleichera oleosa* (Lour.) Oken
Krishna siris (Wheel tree)	*Albizia amara* (Roxb.) Boiv. subsp. *amara*
Kumttha (Gum arabic)	*Acacia senegal* (L.) Willd.
Laurel fig	*Ficus microcarpa* L.f.
Maharukh/Ulloo (Tree of heaven)	*Ailanthus excelsa* Roxb.
Mahua (Honey tree)	*Madhuca longifolia* (Koen.) MacBr. var. *latifolia* (Roxb.) Chev.
Mango	*Mangifera indica* L.
Maulsari (Spanish cherry)	*Mimusops elengi* L.
Moulmein rosewood	*Millettia peguensis* Ali
Nag champa	*Plumeria rubra* var. *acutifolia*
Neeli gulmohur (Jacaranda)	*Jacaranda mimosifolia* D.Don
Neem (Margosa)	*Azadirachta indica* A.Juss.
Peeli gulmohur (Copperpod)	*Peltophorum pterocarpum* (DC.) Back. ex K.Heyne

Pilkhan (White fig)	*Ficus virens* Aiton. var. *virens*
Pink cassia/pink mohur (Java cassia)	*Cassia javanica* L. var. *javanica*
Peepal (Sacred fig)	*Ficus religiosa* L.
Putranjiva (Wild olive)	*Drypetes roxburghii* (Wall.) Hurus.
Quickstick (Mexican lilac)	*Gliricidia sepium* (Jacq.) Kunth ex Walp.
Red cassia (Ceylon senna)	*Cassia roxburghii* DC
Red powder puff (Calliandra)	*Calliandra haematocephala* Hassk.
Ronjh (Distiller's acacia)	*Acacia leucophloea* (Roxb.) Willd.
Royal palm (Bottle palm)	*Roystonea regia* (Kunth) O.F. Cook
Rudraksh tree (Blue fig)	*Elaeocarpus ganitrus* F. Müll.
Sal tree	*Shorea robusta* Gaertn.
Santra (Mandarin orange)	*Citrus reticulata* Blanco
Saptaparni (Devil's tree)	*Alstonia scholaris* (L.) R.Br.
Sausage tree	*Kigelia africana* (Lam.) Benth.
Semal (Red silk cotton)	*Bombax ceiba* L.
Shahtoot (Mulberry)	*Morus alba* L.
Shisham (Indian rosewood)	*Dalbergia sissoo* Roxb. ex DC.
Silver oak (Silky oak)	*Grevillea robusta* A. Cunn. ex R.Br.
Siris (East Indian walnut)	*Albizia lebbeck* (L.) Benth.
Sita-ashok (Sorrowless tree)	*Saraca asoca* (Roxb.) de Wilde
Son champa (Golden champak)	*Michelia champaca* L.
Sonjna (Drumstick tree)	*Moringa oleifera* Lam.
Subabool (Lead tree)	*Leucaena leucocephala* (Lam.) de Wit

Swarna champa	*Plumeria rubra* L. forma *lutea* (Ruiz & Pav.)
Weeping bottlebrush	*Callistemon viminalis* (Soland. ex Gaertn.) G.Don.
White Kanchan (Dwarf bauhinia)	*Bauhinia acuminata* L.
Wild date palm (Sugar date palm)	*Phoenix sylvestris* (L.) Roxb.
Yellow bells	*Tecoma stans* (L.) A.Juss. ex Kunth
Ylang ylang	*Cananga odorata* (Lam.) Hook. f. & Thomson

References

Introduction

1. 'Economic Survey of Delhi 2020-21 (English),' *Planning Department*, Government of NCT of Delhi, https://delhiplanningdelhi.gov.in/planning/economic-survey-delhi-2020-21-english.
2. Singh, Nishant, 'Delhi, The Resilient Phoenix: Destroyed and Rebuilt Seven Times', *Delhi Messenger*, 24 September 2023, https://delhimessenger.in/delhi-the-resilient-phoenix-destroyed-and-rebuilt-seven-times/.
3. Krishen, Pradip, *Trees of Delhi: A Field Guide*, Penguin India DK, Gurugram, 2006, 25.
4. 'Welcome to Delhi Parks and Gardens Society', *Delhi Parks and Gardens Society*, Government of NCT of Delhi, https://dpgs.delhi.gov.in/dpgs/welcome-delhi-parks-and-gardens-society.

January

1. MacDonald, Malcolm, *Birds in My Indian Garden*, Aleph Book Company, New Delhi, 2015, 5.
2. Singh, Khushwant, *Delhi through the Seasons*, HarperCollins India, Gurugram, 2019, 6–7.

3. Agarwal, Priyangi, 'Delhi: Birdcount Returns after Pandemic Break, Some Rare Sightings Delight Enthusiasts', *The Times of India*, 29 November 2021, https://timesofindia.indiatimes.com/city/delhi/birdcount-returns-after-pandemic-break-some-rare-sightings-delight-enthusiasts/articleshow/87968750.cms.
4. 'Delhi: 214 Species Seen on Big Bird Day, Count Lowest in 8 Years', *Hindustan Times*, 25 February 2022, https://www.hindustantimes.com/cities/delhi-news/delhi214-species-seen-on-big-bird-day-count-lowest-in-8-years-101645726267993.html.
5. MacDonald, Malcolm, *Birds in My Indian Garden*, Aleph Book Company, New Delhi, 2015, 10.
6. Ibid. 1.
7. Brother John M. Samaha, S.M., 'Marigolds: Mary's Gold', *All About Mary*, University of Dayton, https://udayton.edu/imri/mary/m/marigolds-marys-gold.php.

February

1. MacDonald, Malcolm, *Birds in My Indian Garden*, Aleph Book Company, New Delhi, 2015, 18.
2. Verma, C.D., 'Signposts Lost in History', *Spectrum*, The Tribune, 10 September 2006, https://www.tribuneindia.com/2006/20060910/spectrum/main2.htm.
3. M. Krishnan, *Of Birds and Birdsong*, Aleph Book Company, New Delhi, 2014, 62.

March

1. Agarwal, Ravi, and Iqbal Malik, 'Walk on the Wild Side', *Ruskin Bond's Green Book*, Ruskin Bond (ed.), Lotus (Roli Books), New Delhi, 2007.
2. Kumar, Nishant, Urvi Gupta, Yadvendradev V. Jhala, Qamar Qureshi, Andrew G. Gosler, and Fabrizio Sergio, 'GPS-Telemetry Unveils the Regular High-elevation Crossing of the Himalayas

by a Migratory Raptor: Implications for Definition of a "Central Asian Flyway"', *Scientific Reports*, Vol. 10, No. 15988, 2020.
3. Krishen, Pradip, *Trees of Delhi: A Field Guide*, Penguin India DK, Gurugram, 2006, 261.
4. Pretor-Pinney, Gavin, *The Cloud Collector's Handbook*, Chronicle Books, San Francisco, CA, 2011, 31.
5. Ali, Salim, *Words for Birds*, Tara Gandhi (ed.), Black Kite (Hachette India), Ranikhet, 2021, 78.
6. Aitken, Edward Hamilton (EHA), *A Naturalist on the Prowl*, Penguin India, Gurugram, 2007, 34.
7. Ali, Salim, *Words for Birds*, Tara Gandhi (ed.), Black Kite (Hachette India), Ranikhet, 2021, 84.
8. Doyle, Arthur Conan, 'The Adventure of the Copper Beeches', *Adventures of Sherlock Holmes*, A.L. Burt Company, New York, 1892, 290.

April

1. Griffiths, Chris, 'Thimmamma Marrimanu: The World's Largest Single Tree Canopy', *BBC*, 20 February 2020, https://www.bbc.com/travel/article/20200219-thimmamma-marrimanu-the-worlds-largest-single-tree-canopy.
2. Roche, Evita, 'The Largest Banyan Tree in the World is Located in India', *Condé Nast Traveller India*, 21 April 2022, https://www.cntraveller.in/story/kolkata-west-bengal-largest-banyan-tree-in-the-world/.
3. Goel, Kapil, 'Rohini Kund at Jagannath Dham, Puri', *ISKCON Desire Tree*, 20 January 2014, https://iskcondesiretree.com/profiles/blogs/rohini-kund-at-jagannath-dham-puri.
4. Krishen, Pradip, *Trees of Delhi: A Field Guide*, Penguin India DK, Gurugram, 2006, 323.
5. Aitken, Edward Hamilton (EHA), *A Naturalist on the Prowl*, Penguin India, Gurugram, 2007, 119.

6. Gaston, A.J., 'Social Behaviour Within Groups of Jungle Babblers (*Turdoides striatus*)', *Animal Behaviour*, Vol. 25, No. 4, 1977, 828–848, https://doi.org/10.1016/0003-3472(77)90036-7.
7. 'Significance of Baisakhi for Sikhs', *The Hindu*, 15 April 2023, https://www.thehindu.com/society/faith/significance-of-baisakhi-for-sikhs/article66737408.ece.
8. Bond, Ruskin, *The Book of Nature*, Penguin Books Limited, Gurugram, 2016,
9. Aitken, Edward Hamilton (EHA), *The Tribes on My Frontier*, Penguin Books India-Viking, Gurugram, 2007, 55.
10. Gooley, Tristan, *The Walker's Guide to Outdoor Clues and Signs*, Hodder Press, London, 2015.
11. Barbano, Paul, 'The Joy Perfume Tree Grows Quite Tall in Its Homeland', *Cape Gazette*, 1 February 2017, https://www.capegazette.com/article/joy-perfume-tree-grows-quite-tall-its-homeland/124731.

May

1. Singh, Khushwant, *Delhi through the Seasons*, HarperCollins India, Gurugram, 2019, 43.
2. Blatter, Ethelbert, and Walter Samuel Millard, *Some Beautiful Indian Trees*, Oxford University Press, Delhi, 1997, 20.
3. Krishen, Pradip, *Trees of Delhi: A Field Guide*, Penguin India DK, Gurugram, 2006, 231.
4. 'The Jacaranda City', *News24*, 9 October 2019, https://showme.co.za/pretoria/tourism/the-jacaranda-city/.
5. 'Ramayana Tours Sri Lanka', *Tourslanka*, https://www.tourslanka.com/excursions/ramayana-in-sri-lanka/.
6. 'Tree Saga - The Sita Ashoka Tree', *Ugaoo*, 22 September 2017, https://www.ugaoo.com/blogs/ornamental-gardening/tree-saga-the-sita-ashoka-tree.
7. Ali, Salim, *The Book of Indian Birds*, OUP India, Delhi, 1997, 237.

8. ESS, 'Bird Life in An Indian Garden', *Ruskin Bond's Green Book*, Ruskin Bond (ed.), Lotus (Roli Books), New Delhi, 2010, 147.
9. NASA Science Editorial Team, 'Super Blood Moon: Your Questions Answered', *The National Aeronautics and Space Administration*, https://science.nasa.gov/solar-system/moon/super-blood-moon-your-questions-answered/.
10. Kelly, Cassie, 'The Tribal Reason Tonight's Full Moon Is the "Flower Moon"', *Inverse*, 10 May 2017, https://www.inverse.com/article/31345-why-is-it-called-the-full-flower-moon.
11. Arnarson, Atli, '6 Science-based Health Benefits of *Moringa oleifera*', *Healthline*, Healthline Media, 6 February 2023, https://www.healthline.com/nutrition/6-benefits-of-moringa-oleifera.
12. Krishen, Pradip, *Trees of Delhi: A Field Guide*, Penguin India DK, Gurugram, 2006, 326.

June

1. Jhabvala, Ruth Prawer, *Heat & Dust*, Harper & Row, Publishers, New York, NY, 1976, 123–124.
2. Haupt, Lyanda Lynn, *Rare Encounters with Ordinary Birds*, Sasquatch Books, Seattle, WA, 2004.
3. Battuta, Ibn, *Travels in Asia and Africa 1325–1354*, Routledge & Kegan Paul Ltd, London, 1929, 367.
4. Muir, John, *John of the Mountains: The Unpublished Journals of John Muir*, Houghton Mifflin Company, Boston, MA, 1938, 438.
5. ESS, 'Bird Life in An Indian Garden', *Ruskin Bond's Green Book*, Ruskin Bond (ed.), Lotus (Roli Books), New Delhi, 2010, 147.
6. Aitken, Edward Hamilton (EHA), *A Naturalist on the Prowl*, Penguin India, Gurugram, 2007, 23.
7. DeWeese, Chris, 'Weather Words: Mackerel Sky', *The Weather Channel*, 3 May 2023, https://weather.com/news/weather/news/2023-05-03-weather-words-mackerel-sky.

8. Gooley, Tristan, 'Weather Lore', *The Natural Navigator*, 22 August 2023, https://www.naturalnavigator.com/the-library/weather-lore/.
9. Swamy, Mahadeswara, 'Kadamba Vriksha: Sri Krishna's Favourite', *Star of Mysore*, 25 August 2019, https://starofmysore.com/kadamba-vriksha-sri-krishnas-favourite/.

July

1. Blatter, Ethelbert, and Walter Samuel Millard, *Some Beautiful Indian Trees*, Oxford University Press, Delhi, 1997, 112.
2. *Kautilya's Arthashastra*, Dr R. Shamasastry (trans.), Government Oriental Library, Mysore, 1951, 127–128.
3. Kalidasa, 'Abhijnanasakuntalam', *The Loom of Time*, Chandra Rajan (trans.), Penguin India, Gurugram, 2006, 176.
4. Krishen, Pradip, *Trees of Delhi: A Field Guide*, Penguin India DK, Gurugram, 2006, 225.
5. Lawrence, D.H., *Lady Chatterley's Lover*, Bantam Books, New York, NY, 1968, 130.
6. Adi Shankara, *Shivananda Lahari*, S. Swaminathan (trans.), *Tamil and Vedas*, 9 July 2012, https://tamilandvedas.com/2012/07/09/four-birds-in-one-sloka-adi-sankara-and-nature/.
7. Kalidasa, 'Meghadutam', *The Loom of Time*, Chandra Rajan (trans.), Penguin India, Gurugram, 2006, 139.

August

1. Agarwal, Simran, 'Western India's Ragamala Paintings Blend Music and Art into a Single Frame', *Scroll.in*, 18 April 2020, https://scroll.in/article/959474/western-indias-ragamala-paintings-blend-music-and-art-into-a-single-frame.
2. Leante, Laura, 'The Cuckoo's Song: Imagery and Movement in Monsoon Ragas', *Monsoon Feelings: A History of Emotions in*

the Rain, Imke Rajamani, Margrit Pernau, and Katherine Butler Schofield (eds.), Niyogi Books, New Delhi, 2018.
3. Sahni, Saumya, '30 Iconic Hindi Songs You Should Listen to When It's Raining Cats & Dogs Outside', *Scoop Whoop*, 29 July 2016, https://www.scoopwhoop.com/music/hindi-rain-songs/.
4. Kalidasa, 'Meghadutam', *The Loom of Time*, Chandra Rajan (trans.), Penguin India, Gurugram, 2006, 141.
5. Sharma, Sunil, 'The Spring of Hindustan: Love and War in the Monsoon in Indo-Persian Poetry', *Monsoon Feelings: A History of Emotions in the Rain*, Imke Rajamani, Margrit Pernau, and Katherine Butler Schofield (eds.), Niyogi Books, New Delhi, 2018, 54.
6. 'The Darkly Veiled June', *Tagore Web*, Kriya Unlimited, https://www.tagoreweb.in/Verses/poems-198/the-darkly-veiled-june-3768.
7. *The Rig Veda*, Ralph T.H. Griffith (trans.), *Sacred Texts*, 1896, https://sacred-texts.com/hin/rigveda/rv05083.htm.
8. Blatter, Ethelbert, and Walter Samuel Millard, *Some Beautiful Indian Trees*, Oxford University Press, Delhi, 1997, 100–102.

September

1. Ali, Salim, *The Book of Indian Birds*, OUP India, Delhi, 1997, 302.
2. Mukherjee, Pulok K., Venkatesan Kumar, N. Satheesh Kumar, and Micheal Heinrich, 'The Ayurvedic Medicine *Clitoria ternatea*—From Traditional Use to Scientific Assessment', *Journal of Ethnopharmacology*, Vol. 120, No. 3, 2008, 291–301, https://doi.org/10.1016/j.jep.2008.09.009.
3. Aitken, Edward Hamilton (EHA), *A Naturalist on the Prowl*, Penguin India, Gurugram, 2007, 41.
4. Ali, Salim, *Words for Birds*, Tara Gandhi (ed.), Black Kite (Hachette India), Ranikhet, 2021, 81.

5. Lohmiller, George, and Becky Lohmiller, 'Dragonflies: Facts, Symbolic Meaning, and Habitat', *Almanac*, Yankee Publishing, 22 February 2024, https://www.almanac.com/content/dragonflies-facts-symbolic-meaning-and-habitat.
6. King James Version. Ecclesiastes 3:2.
7. Bond, Ruskin, *Rain in the Mountains: Notes from the Himalayas*, Penguin India, Gurugram, 2009, 123.
8. Pretor-Pinney, Gavin, *The Cloud Collector's Handbook*, Chronicle Books, San Francisco, CA, 2011, 25.
9. Ridpath, Ian, *Star Tales*, Lutterworth Press, Cambridge, 2018.
10. Krishen, Pradip, *Trees of Delhi: A Field Guide*, Penguin India DK, Gurugram, 2006, 211.

October

1. Montgomery, L.M., *Anne of Green Gables*, Grosset & Dunlap Publishers, New York, NY, 1915, 264.
2. M. Krishnan, *Of Birds and Birdsong*, Aleph Book Company, New Delhi, 2014, 193.
3. Blatter, Ethelbert, and Walter Samuel Millard, *Some Beautiful Indian Trees*, Oxford University Press, Delhi, 1997, 8–9.
4. MacDonald, Malcolm, *Birds in My Indian Garden*, Aleph Book Company, New Delhi, 2015, 217.
5. Dewar, Douglas, *Birds of the Plains*, John Lane The Bodley Head, London, 1909, 125.
6. Phool Walon Ki Sair, https://phoolwaalonkisair.com/history/.
7. Gerard, John, *The Herball*, Adam Islip, Joice Norton and Richard Whitakers, London, 1636.
8. Gabaldon, Diana, *Outlander*, Bantam Doubleday Dell Publishing Group, New York, 1992.
9. M. Krishnan, *Of Birds and Birdsong*, Aleph Book Company, New Delhi, 2014, 65.

November

1. National Audubon Society, https://www.audubon.org/.
2. Bombay Natural History Society, https://bnhs.org/envis.
3. S. Balachandran, Tuhina Katti, and Ranjit Manakadan, *Indian Bird Migration Atlas*, Oxford University Press, Delhi, 2018.
4. Dewar, Douglas, *Indian Birds*, John Lane The Bodley Head, London, 1923, 109.
5. Lawrence, D.H., 'Introduction to *The Dragon of the Apocalypse* by Frederick Carter', *Apocalypse*, Penguin Books, London, 51.
6. Oates, Eugene W., *The Fauna of British India Including Ceylon and Burma, Birds—Vol. I*, Taylor and Francis, London, 1889, 411.

December

1. Kehimkar, Isaac, *The Book of Indian Butterflies*, Bombay Natural History Society, Mumbai, 2008.
2. Aitken, Edward Hamilton (EHA), *The Tribes on My Frontier*, Penguin Books India-Viking, Gurugram, 2007, 110.
3. Singh, Khushwant, *Delhi through the Seasons*, HarperCollins India, Gurugram, 2019, 109–110.

Notes